Gauss and Digital Signal Processing

Volume 3 in the "Scientist and Science" series

Enders Anthony Robinson

Professor Emeritus in the
Maurice Ewing and J. Lamar Worzel Chair
Columbia University in the City of New York

2014

Goose Pond Press

Available from Amazon.com and other retail outlets

Copyright © 2014

by

Enders A. Robinson

Gauss: It is not knowledge but the act of learning,
not possession but the act of getting there,
which grants the greatest enjoyment.

All Rights Reserved Worldwide

Goose Pond Press

Ralph Waldo Emerson told Henry David Thoreau that Goose Pond should be called The Droplet or God's Pond. It was significant to many of Concord's leading literary figures, all of whom walked there often.

Preface

The method of least squares

Adrien-Marie Legendre (1752–1833) made numerous contributions to mathematics. Well-known and important concepts such as the Legendre polynomials and Legendre transformation are named after him. Legendre is also known as the author of *Éléments de Géométrie*, which was published in 1794 and was the leading elementary geometry text for about 100 years. This text rearranged and simplified many of the propositions of Euclid to create a more effective textbook.

In 1806, Legendre published a book on computing the paths of comets. He explained how one could determine the desired parameters of the comet's path. In his calculations, he used *the method of least squares* in order to minimize the experimental error. Legendre's description of the method had an immediate and widespread effect. Today the least squares method has broad applications in linear regression, signal processing, statistics, and curve fitting. The term "least squares method" is a direct translation from the French *méthode des moindres carrés*.

Carl Friedrich Gauss (1777-1855) published his version of the least squares method in 1809, acknowledging that the method first appeared in Legendre's book. However Gauss claimed priority for himself. Gauss went beyond Legendre in both conceptual and technical development, linking the method to probability and providing algorithms for the computation of estimates.

Much of Gauss's development had to wait a long time before finding an appreciative audience, and much was intertwined with others' work, notably Laplace's. Gauss was the leading mathematician of the age, but it was Legendre who crystalized the idea of least squares in a form that was pertinent and applicable.

Gauss assumed that the underlying data followed a bell curve. In time the method became known as Gauss method of least squares and the bell curve became known as the Gaussian curve. The method of least squares as practiced today is not the product of just Legendre and

Gauss, but it is due to many, including George Udny Yule, Norbert Wiener and A. N. Kolmogorov.

The method of least squares is basic in digital signal processing. In many applications it is used to compute the deconvolution operator.

The unit Gaussian curve is

$$y = \left(\frac{1}{\sqrt{2\pi}}\right) \left(e^{-x^2/2}\right)$$

as shown if the figure

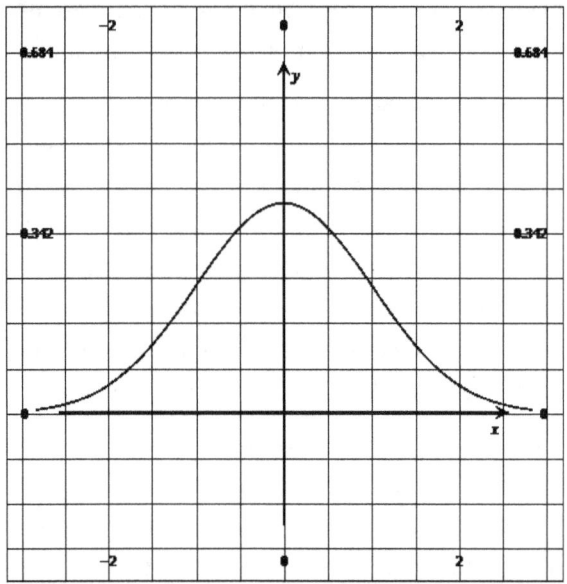

Geological-geophysicists and geophysical-geologists

Under *News of the American Association of Petroleum Geologists (AAPG) 1967* in this bulletin, you will find a timely and thought provoking article by Paul Farren who is well known in Houston Geophysical and Geological circles. Many of you may not know that he has been a member of AAPG since 1948 and is currently serving his second term as District Representative. Paul Farren has been Vice-President of the Society of Exploration Geophysicists (SEG) and past President of the Houston Geophysical Society. He is also a representative on the committee for joint co-operation between AAPG and SEG.

The article by Paul Farren follows.

"I have been asked to discuss the relationship briefly between geologists and geophysicists, especially as it relates to the pending amendment to our geologic constitution to admit "those who have a degree related to geology and have at least seven years' experience in interpreting geophysical evidence into geological terms."

"I attended a most fascinating meeting last Friday night--the Geophysical Society of Houston had the annual Distinguished Lecturer, Dr. Enders Robinson, who gave a talk on deconvolution. Deconvolution sounds like one of those technical terms that doddlebuggers are always fretting over, but I wish every member of HGS had been there to hear Dr. Robinson.

"He set a theme: You need a model before you can have any data processing. The best data processing is that which is done with the best model. Data processing of geophysical evidence to achieve geological ends therefore needs good geological modeling.

"Now, the word deconvolution itself means to unfold. The objective in deconvolving seismic evidence with a computer, then, is to unfold it—from its original heterogeneous complexities into its simpler components. So you, the interpreter, can see what is what. A deconvolved seismic trace is one that is unfolded from its original recorded components so that the interpreter can see which energy is due to multiples, which is due to ghosts, and which is due to the deep Frio, or a fault, for example.

"The best data processing will be done from the best model—one which represents in the best way the geologic box under consideration; and the best final interpretation in the search for a drill site will result from the best recognition of the validity of the geologic model that is revealed after the unfolding.

"The data processing machine has allowed the geophysicist to refocus his attention on the entire process of prediction—to find time structures along seismic traces by unfolding them into their components and seeing how the time events interlock and fit together. The data processing machine by its handling of tremendous masses of information allows man to spend more of his time on the edge of knowledge—applying his unique talents to the field of the unknown. A veritable revolution in the application of man's talents has occurred because of this ability to save time and manipulate masses of information.

"The data processing process even allows man to make of the earth itself a computer which is in turn linked to man's computer—recycling energy into the earth in accordance with the needs determined by sampling of the original energy return from the earth! It is possible for man to see the geophysical and geological results of his own ideas at the time of recording by such use of the earth as a computer, coupled to the new dynamic energy sources. The information from the earth can be unfolded or deconvolved during the recording process itself so that man in his geological thinking can tune man's computer and the earth computer to acquire knowledge of the earth undreamed of a few years ago.

"Refinement of these geophysical tools, refinement of man's talents, refinement of man's geological models to be manipulated by this data processing manipulator is extending the edge of our unknown in tremendous strides. Three dimensional modeling of geological results are in the offing.

"I've simplified Dr. Robinson's wonderfully erudite thinking, of course, but I kept realizing that his objective was beautifully simple: his every effort in applying the most complex geophysical technology of the present and the future was to see the geology a little clearer. Despite what he calls himself, and the tools he uses, he is an active exploration geologist.

"Our President, Mike Halbouty, has gone for broke in an effort to increase effective cooperation between geologist and geophysicists. It has been his thesis from the moment of his acceptance speech in St. Louis. He's heckled me about it, and we have discussed it long and earnestly. We geophysicists appreciate very much his good work, and what we believe to be his good hard headed thinking on this subject. Mike thinks of the petroleum geologists and geophysicists as explorationists, and to heck (not his word!) with the differences—we're all working for the same objective. He's counting on the geologists to be in favor of the carefully worked out amendment involving geophysical memberships. So am I. The vote occurs at Los Angeles.

"The best geologists of the future will be the best geophysicists. The best geophysicists of the future will be the best geologists.

"I wonder how many of us realize that geologists have been using geophysical tools since the advent of the electric logs."

By Paul Farren.

Digital signal processing (DSP) chips

In 1986 the National Academy of Engineering (NAE) of the United States of America elected Enders Robinson as a member with the citation: "For pioneering contributions that have led to the evolution of seismic processing from hand digitization of the 1950s to today's custom deconvolution chip."

The deconvolution chip was the first DSP chip. It served as the prototype of all the succeeding DSP chips.

By 1988 the high-speed digital signal processing (DSP) chip was routinely being incorporated into computer systems. A DSP chip is extremely fast. It is capable of processing a massive quantity of data and producing a response almost instantly. Its speed allows the chip to handle many tasks in the same time. are commonly found in telephone systems and modems. Unless the chip goes extraordinarily fast, it would be of no use. However, its great advantage lies in its capability to do digital signal processing. It can process high quality sound. It can be used for speech synthesis, which allows written data to talk to the user. It enables a computer to accept and store voice mail. It can act as a high-speed modem, sending and receiving two streams of data at once. It can perform as a facsimile machine, transmitting documents along phone lines or computer networks. It can also provide high-speed number crunching. The chip works as an array processor, speeding the performance of graphics for such tasks as three-dimensional modeling.

In brief, the DSP chip makes mobile devices (i.e., hand held computers such as smart phones) versatile and powerful.

DSP chips replace thousands of specialized and costly analog filters. They are found in devices including spy satellites, jet-fighter weapons systems, hearing aids that can be "tuned" to compensate for a wearer's specific hearing loss, stereo recording equipment that restores old recordings to high-fidelity quality, and in data-encrypted secure telephones.

The DSP chip exceeded all expectations. Today, DSP chips are ubiquitous. They are in automobiles to control suspensions that counteract bumps, in stereos to eliminate crackles and pops, and in the helmets of helicopter pilots to cancel the deafening noise made by the rotor blades. Noise cancellation is a particularly valuable feature. Automotive electronic mufflers can cancel noise. The DSP chip can analyze the ambient noise of a factory floor in order to create an opposite sound wave to cancel the noise.

Voice-activated commands and talking computers are commonplace. . DSP's are the intelligence behind automatic automobiles that for safe hands-off driving. DSP chips make complex graphics a powerful tool. An architect can scan around a three-dimensional model. A doctor can rotate images to see all around a patient. The computer can pronounce words, and verbally guide the user. Speech recognition allows the computer takes the place of a stenographer. Digital Signal Processing is no longer in the realm of the mathematician. The DSP chip makes this the technology accessible to all.

Contents

Preface 3
   The method of least squares 3
   Geological-geophysicists and geophysical-geologists........................... 4
   Digital signal processing (DSP) chips 7
Contents 9
Chapter 1. Overview of the seismic method 13
   Basic principles... 13
   Reflection seismology 20
   Mathematical review 22
   Exercises... 26
Chapter 2. Seismic models 29
   Gauss and surfaces... 29
   Ray paths.. 31
   Reverberations... 40
   Layered earth model.. 43
   Einstein addition 50
   Mathematical examples... 54
   Numerical example 55
   Common multiple train.. 58
   Deconvolution.. 59
   Exercises... 65
Chapter 3. Seismic migration 66
   Hologram and wavefront reconstruction 67
   Seismic section and wavefront reconstruction................................. 72
   Wave equation as the basis for seismic migration 77

Chapter 4. Wave motion 79
Introduction 79
Taylor and Fourier 80
Tayler series 82
Pendulum 84
Wave motion 86
Vibrating string 89
Jean Le Rond d'Alembert 93
Wave equation 95
Traveling waves and standing waves 96
Wavefronts and ray paths 98
One-dimensional traveling waves 99
Sinusoidal waves 102
Phase velocity 107
Further points 109
Exercises 112
MIT Whirlwind computer 115

Chapter 5. Hamilton's equations and seismic modeling 116
Wave-particle duality 116
Ray tracing 123
Eikonal equation 131
Ray equations 134
From Hamilton's equations to ray equations 136
Numerical ray tracing 138

Chapter 6. Predictive deconvolution 141
Introduction 141
Least-squares prediction and filtering 143

Least-squares model 145

Matrix formulation 148

Numerical example (1) of least-squares filter 153

Numerical example (2) of least-squares filter 154

Prediction filter 155

Prediction-error filter for unit prediction distance 156

Numerical example of prediction error filter 158

Spiking filter 160

Numerical example of spiking filter 161

Chapter 7. Seismic waves 164

Dual fields 164

Dual sensor 165

d'Alembert equations 170

Einstein deconvolution 172

Chapter 8. Ghost reflections 177

Summary 177

Source ghost and receiver ghost 178

Ghost reflections and reverberations 182

System reflection and transmission coefficients 185

Ghost seismogram 187

Elimination of signature, ghosts, and near-surface multiples 190

Appendix 191

Chapter 9. Fourier series and Fourier transform 194

Periodic functions 194

Jean Baptiste Joseph Fourier 197

General form of Fourier series 201

Complex Fourier series 202

Fourier transform 203

Discrete Fourier transform (DFT) 204

Exercises 205

Chapter 10. Gauss and Maxwell's equations 207

Faraday, Ampere, and Gauss equations 207

Maxwell's equations 209

Chapter 1. Overview of the seismic method

Gauss: We must admit with humility that, while number is purely a product of our minds, space has a reality outside our minds, so that we cannot completely prescribe its properties a priori.

Basic principles

Sedimentary rocks are from sediments deposited under water or laid down on the land by streams or winds. Igneous rocks have solidified from a molten state. Metamorphic rocks were once sedimentary or igneous but were changed by heat and pressure.

A PAY is generally a porous and permeable formation that can produce hydrocarbons. For example, a pay may be a sandstone, or limestone, or hydro-fractured shale, or fissured crystalline rock. Porosity and permeability may be original, dating from sedimentation, or secondary, caused by diagenesis or tectonics. Some pays are tight, and their oil is in fractures and fissures only. Other pays have porosity, but, their permeability being low, they produce petroleum by means of fractures or fissures.

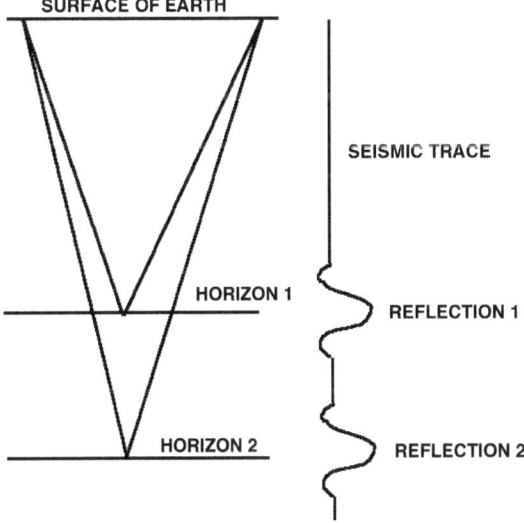

Fig. 1. (left) Ray paths of primary reflections. (right) Corresponding trace.

RAY. A ray is a line showing the direction of flow of radiant energy. See Fig. 1. It is a mathematical device rather than a physical entity.

In practice, one can sometimes produce very narrow beams or pencils, and we might imagine a ray to be the unattainable limit on the narrowness of such a beam.

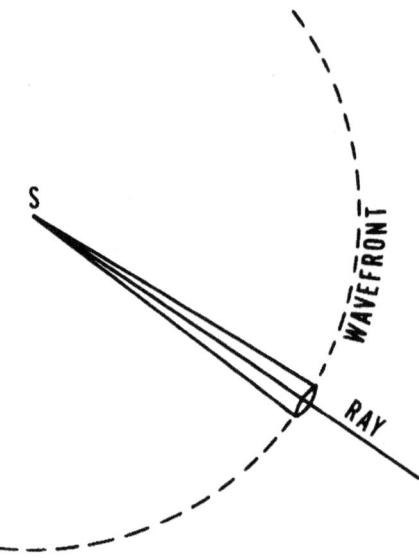

Fig. 2. Ray and wavefront in isotropic medium

In an isotropic medium, i. e., one whose properties are the same in all directions, rays are orthogonal trajectories of the wavefronts. That is to say, they are lines normal (i.e. perpendicular) to the wavefronts at every point of intersection. In such a medium, a ray is parallel to the propagation vector. However, this is not true in anisotropic substances. See Fig. 2.

WAVEFRONT. The dual of the ray is the wavefront. When we look at the waves emanating from the point where we dropped a stone into a still pond, we see the wave motion traveling outward as wavefronts, and the raypaths are only mental constructions. Likewise in seismic interpretation, the rays are only visualized as mathematical abstractions, whereas wavefronts can be actually seen on the seismic sections as reflection events.

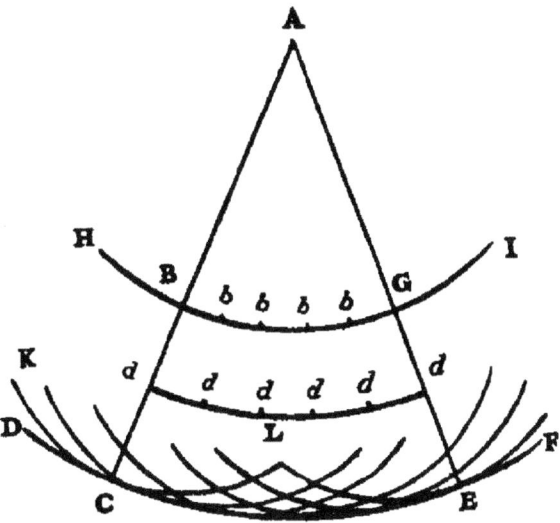

Fig. 3. The depiction of Huygens Principle as given by Huygens

HUYGENS PRINCIPLE. See Fig. 3. Every point on an advancing wavefront is the source of a secondary wavelet. The new wavefront is the envelope of the secondary wavelets. In engineering terms, the secondary wavelet is the impulse response (also called the Green's function) of the wave equation. The output (i.e., the new wavefront) is the spatial convolution of the input (i.e., the original advancing wavefront) with the impulse response (i.e., Green's function).

REFLECTION. See Fig. 4. When an elastic wave strikes an interface separating two media of different physical characteristics, a reflected wave is set up. The reflected wave is present in the same medium as the incident wave. The incident wave carries energy toward the interface, and the reflected wave carries energy away. A refracted wave carries energy into the second medium. The application of Huygens' Principle allows the determination of the angles of reflection and refraction (Snell's law).

$$\frac{\sin\theta_1}{v_1} = \frac{\sin\theta_2}{v_2}$$

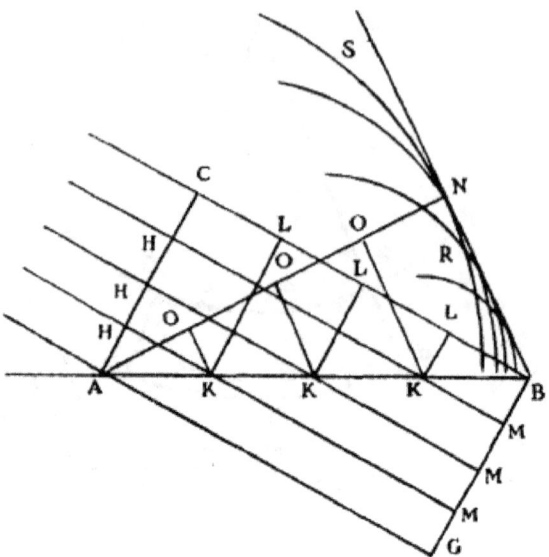

Fig. 4. The depiction of reflection as given by Huygens

REFRACTION. See Fig. 5. The velocities in the two media differ. Waves entering a low-velocity medium experience a shortening of wavelength. The direction of propagation is altered according to Snell's law.

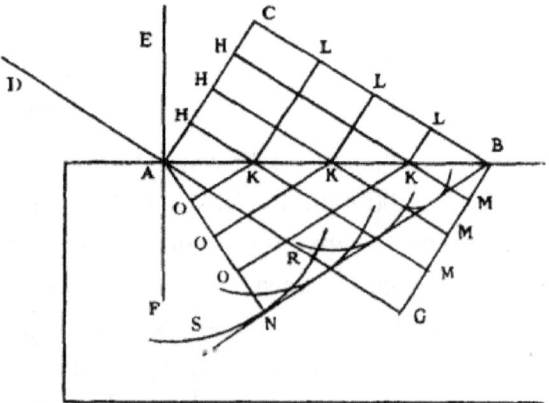

Fig. 5. The depiction of refraction as given by Huygens

WAVES. The distance between crests is the wavelength λ. The number of crests that pass a given point each second is the frequency f. The speed of a crest is the velocity v. Surf at the beach may have $\lambda = 10$ ft, $f = 0.1$ Hz, $v = 1$ ft/s.

Chapter 1. Overview of the seismic method

COMPRESSIONAL WAVES. A compressional wave (or P-wave or longitudinal wave) has the property that, as the wavefront propagates through the medium, the particles of the medium move back and forth in-line with the direction of propagation. The direction of propagation is called the raypath. In an isotropic medium, the raypath is perpendicular to the wavefront. Compressional waves occur in solids, liquids, and gasses.

SHEAR WAVES. In solid material such as the earth, there is a second type of wave that can travel through the body of the material. This wave is called a shear wave (or S-wave or transverse wave). A shear wave the property that its particle motion is perpendicular to the direction of propagation. Air and water can not support shear waves.

SEISMIC COMPRESSIONAL WAVES (also called Primary waves or P-waves). They are longitudinal body waves with particle motion towards and away from the source. Their velocity between 2 km/sec to 14 km/sec depending upon the transmission medium.

SEISMIC SHEAR WAVES. (also called Secondary waves or S-waves). They are transverse body waves with particle motion at right angles to the direction of propagation. They do not travel through liquid. Their velocity is about 0.6 times that of P-waves.

SEISMIC LOVE WAVES (which are surface waves) They are transverse horizontal waves that represent particle motion at right angles to the source direction and in a horizontal plane. Their velocity is between 3 km/sec to 4.5 km/sec.

SEISMIC RAYLEIGH WAVES (which are surface waves) Their particle motion in an elliptical orbit in a vertical plane through the axis of propagation. Their velocity is in the range of 2.5 km/sec to 4 km/sec.

Surface waves are classified as ground roll and considered as noise on exploration data.

POISSON'S RATIO is the ratio of the lateral contraction to the longitudinal extension of an elastic body when it is stretched. Poisson's ratio is an elastic constant that is useful in relating the velocities of the two types of waves, compressional and shear. The expression

$$2(1 - \sigma^2) v_s^2 = v_p^2 (1 - 2\sigma)$$

gives the relationship between the shear velocity V_s, and the compressional velocity V_p, in terms of Poisson's ratio s. Poisson ratios for rocks vary in the range from about 0.2 to about 0.4. The upper bound on the value of Poisson's ratio is 0.5, and at that value V_s is zero, and there are no shear waves. This is the condition that holds for fluids and gases.

ANISOTROPY describes the variation of a physical property with direction. Anisotropy is concerned with directional variation at one point. In contrast, heterogeneity involves variation from point to point. Both anisotropy and heterogeneity depend upon scale, and so the wavelengths involved must be taken into account. Seismic anisotropy refers to the variation of seismic velocity with respect to direction of propagation. In a finely layered medium, the compressional waves may have velocity in the direction of the layering that is up to 20 times higher than the velocity perpendicular to the layering.

Huygens used ellipses to apply his construction to anisotropy. The ellipse represents the velocity function with the minor axis the slow direction, and the major axis the fast direction. The the phase velocity and the ray velocity differ in both direction and magnitude.

POROSITY is defined as pore volume per unit gross volume. Effective porosity is the porosity available to free fluids, excluding unconnected porosity and space occupied by bound water and disseminated shale.

A porous rock consists of a solid skeleton and a continuously connected fluid phase. Intergranular porosity is common in sandstone. Rocks with jugular porosity have small cavities and connected channels created by solution and recrystallization in existing rock. Fracture porosity consists of intersecting cracks caused by tectonic stresses applied to m

DIVING WAVES. The refraction of waves in a strong velocity-gradient zone can reverse the downward direction of the ray and bend it back to the surface.

ABSORPTION is a process in which wave energy is converted to heat during passage through a medium. Absorption is measured by the quality factor Q. Q is defined as 2p times the ratio of the peak energy to the energy dissipated over a cycle. A weathered rock at the surface can have Q as low as 10 or 20; a deep less absorptive rock can have a Q of 200. That is, Q increases and absorption decreases with depth. A Q value of 135 represents an attenuation of 0.2 db per cycle.

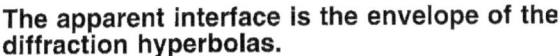

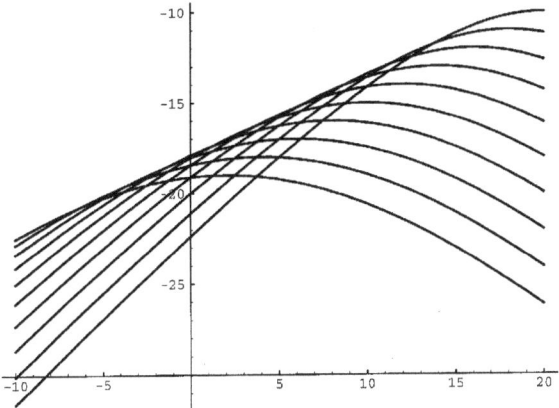

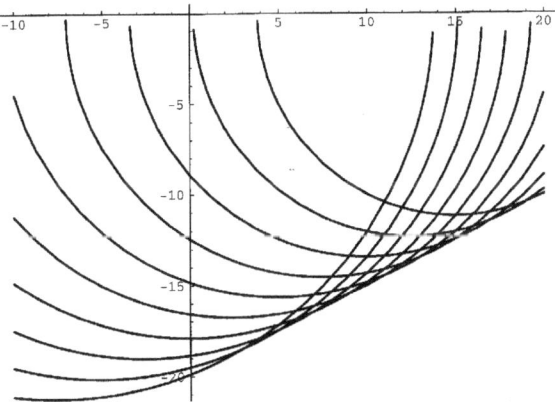

Fig. 6. Apparent interface and true interface

DIFFRACTION. Penetration of wave energy into regions forbidden by geometric optics, e.g., the bending of wave energy around obstacles without obeying Snell's law. Diffraction allows waves to travel in the geometric shadow of a barrier. See Fig. 6.

FRESNEL ZONE. For coincident source and receiver, the first Fresnel zone has radius R_1. The annular ring is the second Fresnel zone. l is the dominant wavelength. Most of the energy reflected from the first Fresnel zone interferes constructively, but the outer portion does not contribute much.

Reflection seismology

Oil and natural gas are found in traps located deep within the sedimentary layers of the earth's crust. The reflection seismic method is by far the most important geophysical technique used to map the underground interfaces between the geological strata. These seismic maps together with other geological information are used to locate favorable drilling sites for exploratory wells. The seismic method involves the transmission of man-made seismic disturbances into the earth, and the recording of the resulting echoes (or reflections) from the geologic interfaces. These recordings (or traces) are in the form of digital signals, one digital signal for each source and receiver pair. Various types of noise interfere with the desired reflections, and so the interpretation of the raw data recorded presents a formidable problem. A whole branch of geophysics, called geophysical digital signal processing, has grown up since 1952 to provide the people, equipment, and methods to analyze these digital signals.

There are two sequential operations, which come under the headings of the deconvolution step and the re-convolution step. Deconvolution removes the unwanted multiple reflections and other interference from the raw data in order to produce sparse signals. Re-convolution converts the sparse signals into holograms of the underground structure by means of Huygens construction and other wave-equation methods. In seismic processing, the re-convolution step is called migration.

Reflection seismology is a method of mapping the subsurface structure of the earth from knowledge of the arrival times of events reflected from the subsurface layers. The earth's sedimentary layers are approximately horizontal, but they do have features such as anticlines, unconformities, and faults that can serve as traps for petroleum. See Fig. 7.

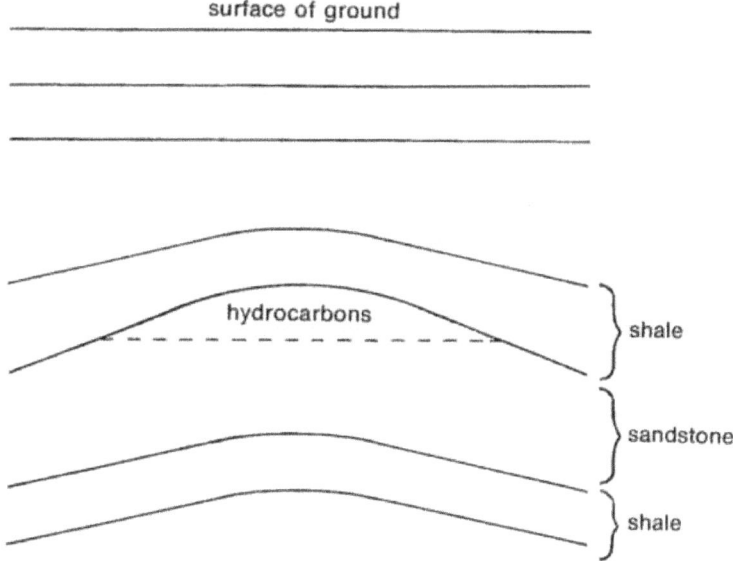

Fig. 7. Anticline serving as hydrocarbon trap.

In order to map the subsurface, the geophysicist must convert the received seismic traces which record the events as a function of time into a function of depth. That is, digital signal recorded at the surface must be transformed into a depth section of the earth. Unlike radio waves, seismic waves have a velocity which is very much dependent on the medium. Thus, the velocity changes as the waves travel into the earth. Generally the velocity increases with depth, although occasionally there may be layers in which a decrease in velocity occurs. For a given surface point, the velocity plotted as a function of depth is called the velocity function. Thus, in reflection seismology there are two equally important variables, time of reflected events and velocity. From knowledge of these variables the depth to the reflecting horizons must be determined. Because there are important lateral changes in velocity, that is, because the velocity function varies from one location to another, a given velocity function cannot be assumed to be valid for an entire prospect. As a result, the velocity function must be continually corrected from place to place over the area of exploration.

One method of measuring the velocity function is to drill a deep hole, namely an oil well, and determine the velocity by placing seismic detectors in the hole at various depths. However, in most cases, it is necessary to estimate the velocity function by measurements confined to the surface, since oil wells are available only in old prospects. The velocity function can be estimated by considering the time differentials of the same event sampled by a lateral array of sources and receivers. Any such estimate always depends upon a ceteris paribus (other things being equal) assumption. In brief, data recorded as a function of the two surface dimensions source position and receiver position as well as the time dimension t are used to compute a velocity function. By means of this velocity function, the data are transformed into a cross section of the earth, given by a surface dimension x and a depth dimension z. The three operations that perform this transformation are deconvolution, stacking, and migration.

Mathematical review

The essence of mathematics is not to make simple things complicated, but to make complicated things simple. The subject matter of digital signal processing is concerned with massive arrays of numbers. Matrix theory is a way of handling arrays.

A matrix is a rectangular grid of numbers. The matrix

$$\begin{pmatrix} 1 & 2 & 3 \\ 4 & 5 & 6 \end{pmatrix}$$

has 2 rows and 3 columns. It is called a 2×3 matrix.

A 2×2 matrix is called a square matrix. We can do some special things with square matrices that we cannot do with other sizes. For example, we can only take determinants of square matrices. We can only take inverses of square matrices with non-zero determinants.

A $1 \times n$ matrix is called a row vector. A $m \times 1$ matrix is called a column vector. We can do things with vectors. For example, we can multiply a $1 \times n$ row vector with a $n \times 1$ column vector. The result is a single number called the dot product. For example,

$$(1 \quad 2 \quad 3) \begin{pmatrix} 4 \\ 5 \\ 6 \end{pmatrix} = 1 \times 4 + 2 \times 5 + 3 \times 6 = 4 + 10 + 18 = 30$$

The dot product is 30.

Chapter 1. Overview of the seismic method

In order to refer to a specific entry, we use a special tagging system. It is based upon rows and columns. The entry a in row i and column j is denoted by a_{ij}.

In multiplying two matrices, you need to be careful. First of all, the size of the two matrices must be compatible. Say the size of matrix A is 3×4 and the size of matrix B is 4×1. Then the product has size 3×1. In general, if the size of matrix A is $m \times n$ and the size of matrix B is $n \times p$, then the size of the product $C = AB$ is $m \times p$. The entry c_{1j} is the dot product of row i of matrix A and column j of matrix B.

A digital signal may be represented either by a row vector or by a column vector. In this book we will generally use a column vector. For example, the digital signal might be

$$a = \begin{pmatrix} 3 \\ 2 \\ 1 \end{pmatrix}$$

The corresponding set is the matrix made up of the digital signal and its delayed versions; for example,

$$A = \begin{pmatrix} 3 & 0 & 0 & 0 \\ 2 & 3 & 0 & 0 \\ 1 & 2 & 3 & 0 \\ 0 & 1 & 2 & 3 \\ 0 & 0 & 1 & 2 \\ 0 & 0 & 0 & 1 \end{pmatrix}$$

Let the other signal be

$$b = \begin{pmatrix} 4 \\ 5 \\ 6 \\ 7 \end{pmatrix}$$

The convolution of the two signals is

$$Ab = \begin{pmatrix} 3 & 0 & 0 & 0 \\ 2 & 3 & 0 & 0 \\ 1 & 2 & 3 & 0 \\ 0 & 1 & 2 & 3 \\ 0 & 0 & 1 & 2 \\ 0 & 0 & 0 & 1 \end{pmatrix} \begin{pmatrix} 4 \\ 5 \\ 6 \\ 7 \end{pmatrix} = \begin{pmatrix} 12 \\ 23 \\ 32 \\ 38 \\ 20 \\ 7 \end{pmatrix}$$

GAUSS AND DIGITAL SIGNAL PROCESSING

Unfortunately, the matrix A is not square, so it cannot be inverted. We will now disclose a special method. Let us define a new matrix that is square.

$$A = \begin{pmatrix} 3 & 0 & 0 & 0 & 0 & 0 \\ 2 & 3 & 0 & 0 & 0 & 0 \\ 1 & 2 & 3 & 0 & 0 & 0 \\ 0 & 1 & 2 & 3 & 0 & 0 \\ 0 & 0 & 1 & 2 & 3 & 0 \\ 0 & 0 & 0 & 1 & 2 & 3 \end{pmatrix}$$

We now let

$$b = \begin{pmatrix} 4 \\ 5 \\ 6 \\ 7 \\ 0 \\ 0 \end{pmatrix}$$

The convolution of the two signals is now

$$c = Ab = \begin{pmatrix} 3 & 0 & 0 & 0 & 0 & 0 \\ 2 & 3 & 0 & 0 & 0 & 0 \\ 1 & 2 & 3 & 0 & 0 & 0 \\ 0 & 1 & 2 & 3 & 0 & 0 \\ 0 & 0 & 1 & 2 & 3 & 0 \\ 0 & 0 & 0 & 1 & 2 & 3 \end{pmatrix} \begin{pmatrix} 4 \\ 5 \\ 6 \\ 7 \\ 0 \\ 0 \end{pmatrix} = \begin{pmatrix} 12 \\ 23 \\ 32 \\ 38 \\ 20 \\ 7 \end{pmatrix}$$

The new matrix gives the same result as before. In other words, we convolved b with a to obtain c.

We can invert the matrix A to obtain

$$A^{-1} = \begin{pmatrix} 3 & 0 & 0 & 0 & 0 & 0 \\ 2 & 3 & 0 & 0 & 0 & 0 \\ 1 & 2 & 3 & 0 & 0 & 0 \\ 0 & 1 & 2 & 3 & 0 & 0 \\ 0 & 0 & 1 & 2 & 3 & 0 \\ 0 & 0 & 0 & 1 & 2 & 3 \end{pmatrix}^{-1} = \begin{pmatrix} \frac{1}{3} & 0 & 0 & 0 & 0 & 0 \\ -\frac{2}{9} & \frac{1}{3} & 0 & 0 & 0 & 0 \\ \frac{1}{27} & -\frac{2}{9} & \frac{1}{3} & 0 & 0 & 0 \\ \frac{4}{81} & \frac{1}{27} & -\frac{2}{9} & \frac{1}{3} & 0 & 0 \\ -\frac{11}{243} & \frac{4}{81} & \frac{1}{27} & -\frac{2}{9} & \frac{1}{3} & 0 \\ \frac{10}{729} & -\frac{11}{243} & \frac{4}{81} & \frac{1}{27} & -\frac{2}{9} & \frac{1}{3} \end{pmatrix}$$

Chapter 1. Overview of the seismic method

The process of deconvolution undoes the result of convolution. We have

$$A^{-1}c = \begin{pmatrix} \frac{1}{3} & 0 & 0 & 0 & 0 & 0 \\ -\frac{2}{9} & \frac{1}{3} & 0 & 0 & 0 & 0 \\ \frac{1}{27} & -\frac{2}{9} & \frac{1}{3} & 0 & 0 & 0 \\ \frac{4}{81} & \frac{1}{27} & -\frac{2}{9} & \frac{1}{3} & 0 & 0 \\ -\frac{11}{243} & \frac{4}{81} & \frac{1}{27} & -\frac{2}{9} & \frac{1}{3} & 0 \\ \frac{10}{729} & -\frac{11}{243} & \frac{4}{81} & \frac{1}{27} & -\frac{2}{9} & \frac{1}{3} \end{pmatrix} \begin{pmatrix} 12 \\ 23 \\ 32 \\ 38 \\ 20 \\ 7 \end{pmatrix} = \begin{pmatrix} 4 \\ 5 \\ 6 \\ 7 \\ 0 \\ 0 \end{pmatrix} = b$$

In other words, we deconvolved c with a to obtain b. The first column of A^{-1}; namely

$$\begin{pmatrix} \frac{1}{3} \\ -\frac{2}{9} \\ \frac{1}{27} \\ \frac{4}{81} \\ -\frac{11}{243} \\ \frac{10}{729} \end{pmatrix}$$

is the approximate inverse of the signal a. In other words, we have

$$A^{-1}a = \begin{pmatrix} \frac{1}{3} & 0 & 0 & 0 & 0 & 0 \\ -\frac{2}{9} & \frac{1}{3} & 0 & 0 & 0 & 0 \\ \frac{1}{27} & -\frac{2}{9} & \frac{1}{3} & 0 & 0 & 0 \\ \frac{4}{81} & \frac{1}{27} & -\frac{2}{9} & \frac{1}{3} & 0 & 0 \\ -\frac{11}{243} & \frac{4}{81} & \frac{1}{27} & -\frac{2}{9} & \frac{1}{3} & 0 \\ \frac{10}{729} & -\frac{11}{243} & \frac{4}{81} & \frac{1}{27} & -\frac{2}{9} & \frac{1}{3} \end{pmatrix} \begin{pmatrix} 3 \\ 2 \\ 1 \\ 0 \\ 0 \\ 0 \end{pmatrix} = \begin{pmatrix} 1 \\ 0 \\ 0 \\ 0 \\ 0 \\ 0 \end{pmatrix}$$

Exercises

1. Determine the unit impulse response b of a causal linear time-invariant system from the observed causal input a and causal output c.

a	1	0	0
c	1	-2	0

$$A = \begin{pmatrix} 3 & 0 & 0 \\ -2 & 3 & 0 \\ 0 & -2 & 3 \end{pmatrix} \text{ and } c = \begin{pmatrix} 1 \\ 0 \\ 0 \end{pmatrix}$$

The answer is b where

$$A^{-1}c = b$$

is given by

$$\begin{pmatrix} 3 & 0 & 0 \\ -2 & 3 & 0 \\ 0 & -2 & 3 \end{pmatrix}^{-1} \begin{pmatrix} 1 \\ 0 \\ 0 \end{pmatrix} = \begin{pmatrix} \frac{1}{3} \\ \frac{2}{9} \\ \frac{4}{27} \end{pmatrix}$$

As a check, we have

$$Ab = \begin{pmatrix} 3 & 0 & 0 \\ -2 & 3 & 0 \\ 0 & -2 & 3 \end{pmatrix} \begin{pmatrix} \frac{1}{3} \\ \frac{2}{9} \\ \frac{4}{27} \end{pmatrix} = \begin{pmatrix} 1 \\ 0 \\ 0 \end{pmatrix} = c$$

2. Determine the unit impulse response b of a causal linear time-invariant system from the observed causal input a and causal output c.

a	3	-2	0
c	1	-2	0

$$A = \begin{pmatrix} 3 & 0 & 0 \\ -2 & 3 & 0 \\ 0 & -2 & 3 \end{pmatrix} \text{ and } c = \begin{pmatrix} 1 \\ -2 \\ 0 \end{pmatrix}$$

The answer is **b** where

$$A^{-1}c = b$$

is given by

$$\begin{pmatrix} 3 & 0 & 0 \\ -2 & 3 & 0 \\ 0 & -2 & 3 \end{pmatrix}^{-1} \begin{pmatrix} 1 \\ -2 \\ 0 \end{pmatrix} = \begin{pmatrix} \frac{1}{3} \\ -\frac{4}{9} \\ -\frac{8}{27} \end{pmatrix}$$

As a check, we have

$$Ab = \begin{pmatrix} 3 & 0 & 0 \\ -2 & 3 & 0 \\ 0 & -2 & 3 \end{pmatrix} \begin{pmatrix} \frac{1}{3} \\ -\frac{4}{9} \\ -\frac{8}{27} \end{pmatrix} = \begin{pmatrix} 1 \\ -2 \\ 0 \end{pmatrix} = c$$

3. Determine the unit impulse response **b** of a causal linear time-invariant system from the observed causal input **a** and causal output **c**.

a	3	−2	1	0
c	1	−2	−1	2

$$A = \begin{pmatrix} 3 & 0 & 0 & 0 & 0 & 0 \\ -2 & 3 & 0 & 0 & 0 & 0 \\ 1 & -2 & 3 & 0 & 0 & 0 \\ 0 & 1 & -2 & 3 & 0 & 0 \\ 0 & 0 & 1 & -2 & 3 & 0 \\ 0 & 0 & 0 & 1 & -2 & 3 \end{pmatrix} \quad \text{and} \quad c = \begin{pmatrix} 1 \\ -2 \\ -1 \\ 2 \\ 0 \\ 0 \end{pmatrix}$$

The answer is **b** where

$$A^{-1}c = b$$

is given by

$$\begin{pmatrix} 3 & 0 & 0 & 0 & 0 & 0 \\ -2 & 3 & 0 & 0 & 0 & 0 \\ 1 & -2 & 3 & 0 & 0 & 0 \\ 0 & 1 & -2 & 3 & 0 & 0 \\ 0 & 0 & 1 & -2 & 3 & 0 \\ 0 & 0 & 0 & 1 & -2 & 3 \end{pmatrix}^{-1} \begin{pmatrix} 1 \\ -2 \\ -1 \\ 2 \\ 0 \\ 0 \end{pmatrix} = \begin{pmatrix} \dfrac{1}{3} \\ -\dfrac{4}{9} \\ -\dfrac{20}{27} \\ \dfrac{26}{81} \\ \dfrac{112}{243} \\ \dfrac{146}{729} \end{pmatrix}$$

As a check, we have

$$\boldsymbol{Ab} = \begin{pmatrix} 3 & 0 & 0 & 0 & 0 & 0 \\ -2 & 3 & 0 & 0 & 0 & 0 \\ 1 & -2 & 3 & 0 & 0 & 0 \\ 0 & 1 & -2 & 3 & 0 & 0 \\ 0 & 0 & 1 & -2 & 3 & 0 \\ 0 & 0 & 0 & 1 & -2 & 3 \end{pmatrix} \begin{pmatrix} \dfrac{1}{3} \\ -\dfrac{4}{9} \\ -\dfrac{20}{27} \\ \dfrac{26}{81} \\ \dfrac{112}{243} \\ \dfrac{146}{729} \end{pmatrix} = \begin{pmatrix} 1 \\ -2 \\ -1 \\ 2 \\ 0 \\ 0 \end{pmatrix} = \boldsymbol{c}$$

Chapter 2. Seismic models

Gauss: Like a sudden flash of lightning, the riddle was solved. I am unable to say what the conducting thread was that connected what I previously knew with what made my success possible.

Gauss and surfaces

Carl Friedrich Gauss was director of the astronomical Observatory at Göttingen from 1807 to his death in 1855. One of the greatest scientists of all time, his work combined a fertility of concepts with the originality of methods.

For Gauss, theory and practice were intimately related. He approached the theory of surfaces primarily as a result of his work on triangulation, where the emphasis is on measurements between points on the surface of the earth. Consequently he saw the surface not so much the boundary of a solid body, but as a film, a two-dimensional entity not necessarily attached to a three-dimensional body. A piece of such a surface can be bent and we can ask the properties of the film which do not change under bending. A two-dimensional alien being, confined to the two=dimensional surface, would be unaware of any outside space. However, the alien being would be able to find the path of the shortest distance between the two points, where the length of the path would be measured on the two-dimensional surface. The alien being would also be able to measure the angle between two directions on the surface. The length and angle in question are intrinsic properties of the two-dimensional surface. Gauss drew from his work as a surveyor the inspiration for his profound reappraisal of the general theory of surfaces.

The fields of his activity ranged from the theory of numbers to complex variable theory, and from celestial mechanics to the electric telegraph, of which he was one of the inventors. His fundamental work of surface theory was, like much of Gauss's work, were written in Latin. Av English translation exists with the title General investigations on curved surfaces. Gauss published other works on questions pertaining to geometry such as his paper on conference on conformal representation in 1822, but he kept some of his boldest Ideas to himself, notably his ideas on non-Euclidean geometry.

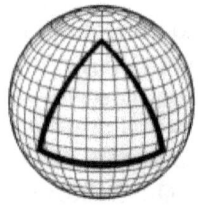 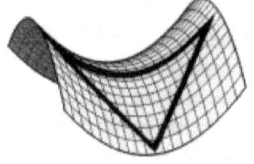 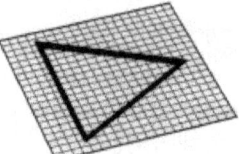

Positive Curvature   Negative Curvature   Flat Curvature

A smooth surface has two principal curvatures. The Gaussian curvature is a product of the two principal curvatures. A positive Gaussian curvature value means the surface is bowl-like. A negative value means the surface is saddle-like. A zero value means the surface is flat in at least one direction. (Planes, cylinders, and cones have zero Gaussian curvature).

An important theorem of Gauss is the one that that he called *Theorema Egregium* which means *most excellent theorem*. It states that the Gaussian curvature of the surface is a bending invariant. When we call the Gaussian curvature a bending invariant, we mean that it is unchanged by such deformations of the surface that do not involve stretching, shrinking, or tearing. Such bending leaves the distance between two points on the service, as measured along a curve on the surface, unchanged. Such bending also leaves the angle between two tangents at a point unchanged. An example of this bending can be obtained by deforming a piece of paper without changing its elastic properties. In other words, we should not get the paper wet A curve on the piece of paper keeps the same length when the paper is bent into some other shape.

The only closed service of constant positive curvature without singularities is the sphere. A sphere cannot be bent. A sphere with a hole in it can be bent.

Surfaces of zero Gaussian curvature are developable surfaces. The simplest example the surface of constant nonzero Gaussian curvature is a sphere. Its curvature is a positive constant. However, the sphere is not the only surface with this property. Take the cap of a sphere and let it consist of some inelastic material such as thin brass. We can give this piece of brass all kinds of shapes that retain the same constant curvature.

Ray paths

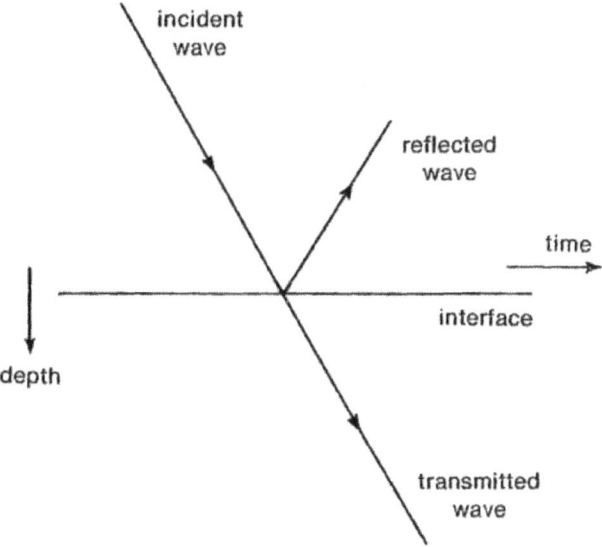

Fig. 1. Reflection and transmission at an interface. (Vertical travel paths are assumed, but the lines are given a horizontal displacement to indicate the passage of time.)

Let us first discuss deconvolution. Basic to the understanding of deconvolution in the processing of reflection seismic data is the development of a model of the earth consisting of a finite number of layered strata. With this approach in mind, let us look at a single horizontal interface between two sedimentary layers. Fig. 1 shows how an interface divides an incident wave into a reflected wave and a transmitted wave.

We assume that there is no loss due to absorption. When a wave of unit energy strikes the interface, some of the energy is reflected and the remainder of the energy is transmitted. The amount of reflected energy depends upon the reflection coefficient of the interface. The reflection coefficient is a number that lies between minus one and plus one. The reflection coefficient represents the information we wish to know about the interface.

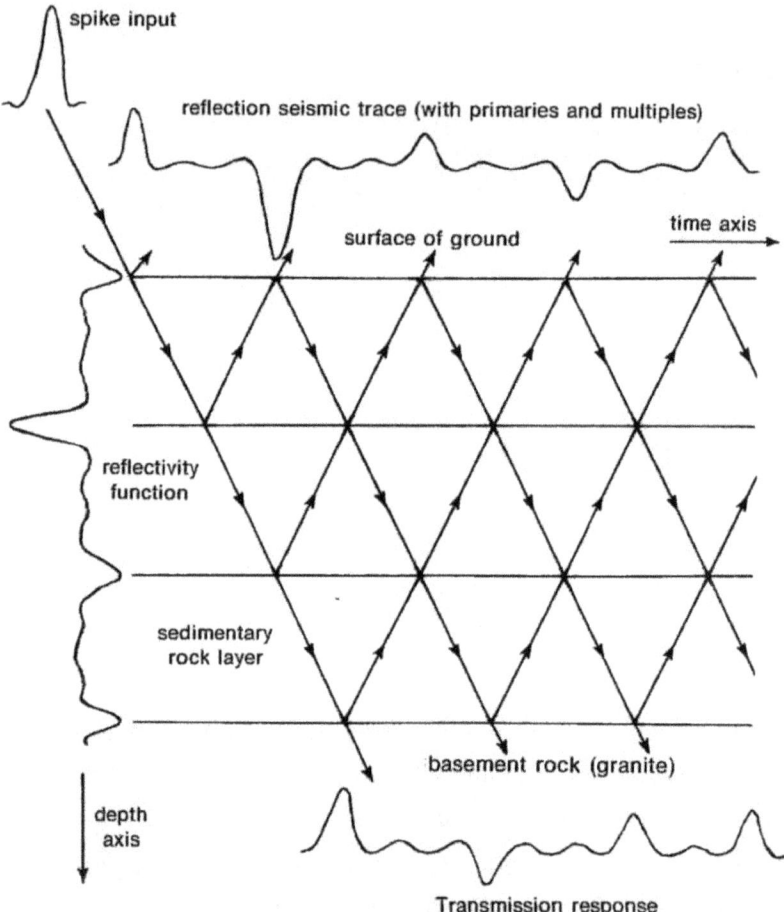

Fig. 2. (left) reflectivity function; (top) reflection response; (i.e. seismic trace) and (bottom) transmission response

See Fig. 2. The sequence of reflection coefficients for all the interfaces is called the reflectivity function, and represents the desired information. The reflectivity function determines the reflection response (i.e. seismic trace) and the transmission response (i.e. the waveform which is transmitted into the granite and never returned to the layered system).

Chapter 2. Seismic models

A seismic source (such as an explosion) sets off a sharp impulsive wave or spike traveling down into the earth. At each interface reflection and transmission occur. Thus there are many wave paths as the waves travel back and forth between the subsurface interfaces. Eventually all the energy must either return to the surface (which gives the reflection response) or be transmitted down into the basement rock (which gives the transmission response).

Receivers on the surface record the reflection response. The reflection response is the digital signal commonly called a seismic trace. The horizontal lines show the interfaces, and the diamond shaped grid shows the tracks of the various ray paths. In this discussion we are considering only vertically traveling waves, but the ray paths are given a horizontal displacement to indicate the passage of time. In the left hand column we see the reflectivity function which shows a pip at each interface. In the upper left hand corner we see the spike input. This input energy eventually comes out as the reflection seismic trace (the curve at the top) and the transmission response (the curve at the bottom). The seismic trace represents the known information.

Two types of events occur on the observed reflection seismogram, namely the primary events and the multiple events. The primary events represent energy that has travelled a direct path from surface to reflecting interface and then back to the surface. However for every primary event there are many multiple events. A multiple event is one that makes one or more round-trip paths within the sedimentary layers before returning to the surface of the earth. A primary path together with various multiple paths can produce reflected energy on the seismogram at the same arrival time. Thus energy from both types of paths can combine to make up an observed event on the reflection seismogram. In a sense we may say that certain multiple energy (which actually never reaches an interface) reinforces the primary event due to the interface. In this sense multiples are good in that they can reinforce primary reflections.

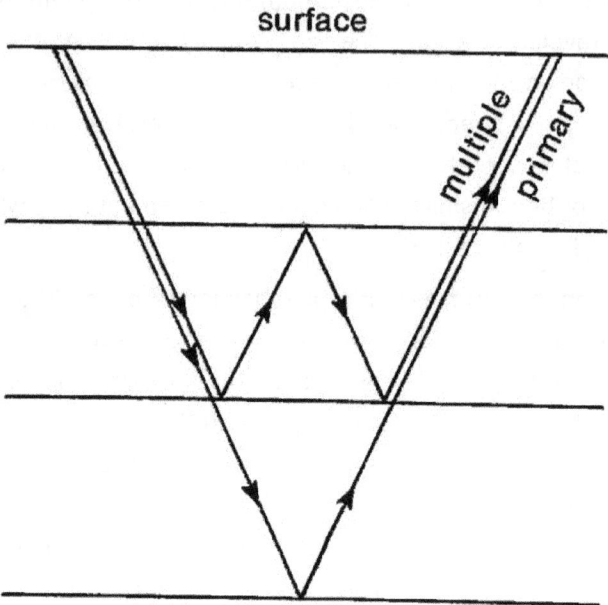

Fig. 3. Reinforcement of a primary with a multiple

A typical reinforcement is shown in Fig. 3. Why is such reinforcement necessary? As we have seen, each time seismic energy is transmitted through an interface some energy is lost to reflection. Thus, for a primary event, the amplitude of the source pulse must be multiplied by each layer's transmission coefficient in the direct path down to a reflecting horizon, as well as by each layer's transmission coefficient in the direct path back to the surface. Each of these two-way transmission coefficients in magnitude is less than one, so the net effect is that the transmission losses can greatly reduce the amplitude of a primary event. Reinforcement of primary events by multiples is nothing more than the regaining of the primary of some of the energy that was lost by the transmission effects.

We have given the advantageous effects of multiples. Now let us turn to the disadvantageous effects. The great disadvantage is that a multiple event can appear on a seismic trace where no primary event exists. Thus when we see such an event, without further analysis, we would mistake it for a primary event. More generally, many such multiples would interfere with the primaries, masking them and making it impossible to delineate them. Thus multiples represent a serious kind of signal-generated noise that hinders an accurate interpretation of seismic events.

There are two main ways of reducing the disadvantageous effects of multiple events in current seismic processing practice. These two processing methods are stacking and deconvolution. Stacking works in the space domain whereas deconvolution works in the time domain. Stacking is a method of averaging over space that reinforces the primaries and cancels the multiples. Deconvolution is a method of averaging over time that does the same thing. Ideally, stacking and deconvolution would be carried out by means of overall multi-channel space-time deconvolution operators. From a practical point of view these processing operations are carried out separately in such a manner that the overall seismic data processing package is extremely robust and stable. In this section we discuss the deconvolution operation as a separate entity, distinct from the other processing methods.

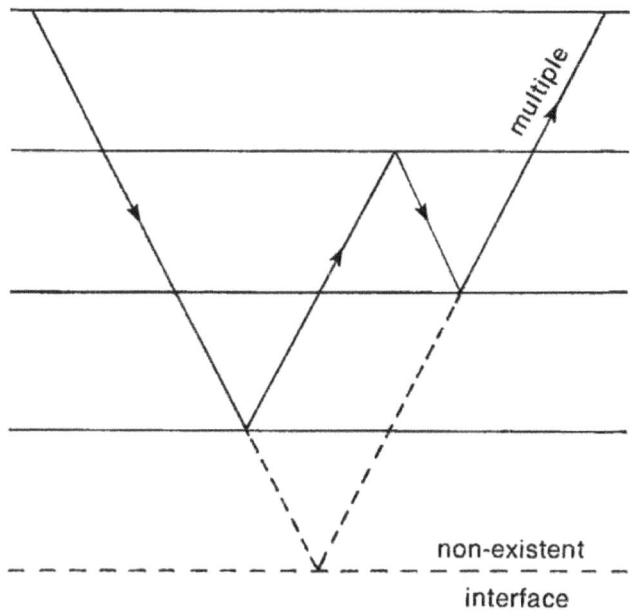

Fig. 4. The multiple (solid path) may be mistaken as a primary (dashed path). If a multiple is so mistaken, the result would be that a non-existent interface is incorrectly inferred.

As we have seen, two distinct possibilities can occur. One is that a multiple event can constructively reinforce a primary event, as shown in Fig. 3. In Fig. 4 we see another possibility, which we want to avoid. This other possibility is that a multiple event can confound primary events. Is there an example in nature where multiples only behave constructively and not destructively? The answer is yes, and we will soon describe this optimal situation.

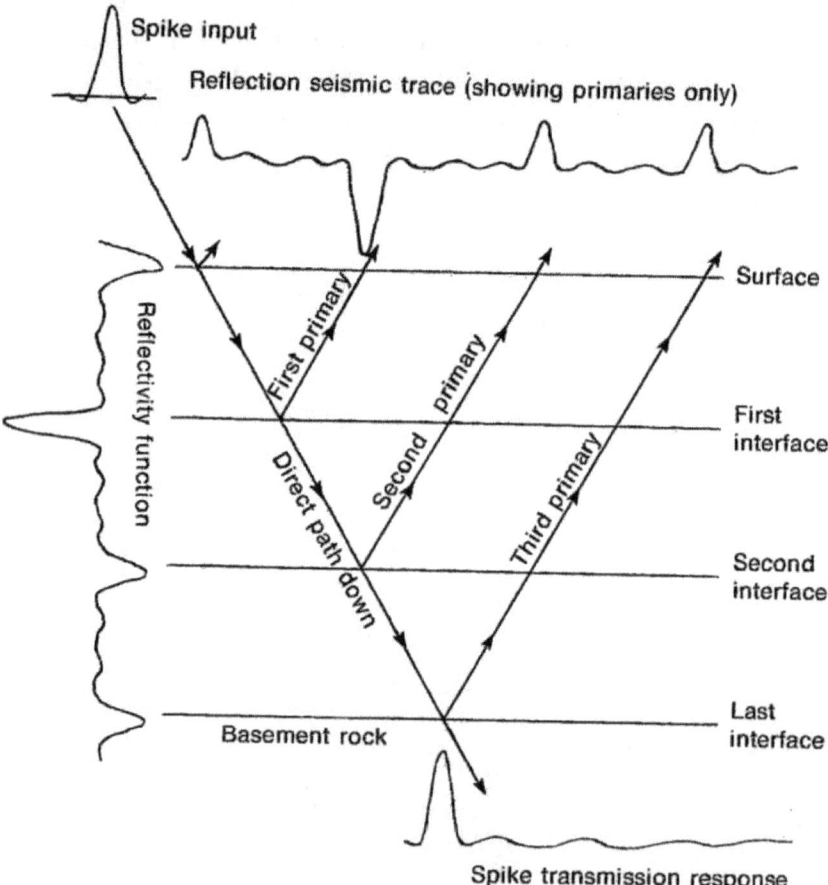

Fig. 5. If there were no multiples and no transmission losses, then the seismic trace would be the same as the reflectivity function, and the transmission response would be the same as the input.

An oil well drilled in a sedimentary basin will reveal the layering. First we will describe a hypothetical (but not physically realizable) situation

Chapter 2. Seismic models

as illustrated in Fig. 5. More precisely, for this example we assume that there are no transmission losses on passage though the interfaces and that there are no multiple reflections.

If we plot the reflection coefficients of these layers as a function of two-way travel-time we obtain the so-called reflectivity function. Ideally in the case of distinct well-defined layers, this reflectivity function would have a pip at each interface. The size of such a pip would be equal to the reflection coefficient at that interface. It is an observable fact that the magnitudes of most of the reflection coefficients encountered are small, in fact, much less than one in magnitude. The large reflection coefficients are associated with major interfaces. In the hypothetical case of no multiples and no transmission losses, the reflection seismogram would be the result obtained by attaching the source spike to each pip on the reflectivity function. Thus for this hypothetical (but not physically realizable) situation the reflection seismogram would be the reflectivity function itself. Also the transmission response would be the same as the input, namely a spike.

In any physical situation, there must be transmission losses and multiple reflections. The effect of transmission losses is to diminish the magnitudes of the amplitudes of the primaries from what the primaries would be without transmission losses. The effect of multiples is to add many more events to the seismic trace from what the trace would be with primaries only. If there were no mutual cancelling of effects, then the seismic trace would no longer look like the reflectivity function. Also the transmission response would no longer look like the input.

The optimal situation which we mentioned above can now be described. In the best possible case we would want the reflection seismogram to look exactly like the reflectivity function. In this optimal situation, the energy in the primary events lost through the transmission effects would be exactly counterbalanced by the energy gained through the multiples. That is, transmission effects and multiple effects would exactly cancel each other, and we would be left with the ideal seismogram, namely the reflectivity function. Also, in this optimal situation, the transmission response would look like the spike input. Can this optimal situation be attained in nature? The answer is that such an optimal situation can actually exist in nature, and in face approximations to it are not uncommon.

The optimal situation arises when the reflectivity function is a white noise sequence: that is, when the sequence of reflection coefficients is the same as a sequence of statistically uncorrelated random observations. As it is well known, a white noise sequence has the property that its autocorrelation function is a spike. Thus if we compute the autocorrelation function of a reflectivity function, and find that this autocorrelation is a spike, then we know that the reflection seismogram (in the case of a spike source) will look the same as the reflectivity function.

This optimal situation, or at least various approximations to it, actually occurs in particular geographic areas of the world where early oil prospecting has taken place. The actual seismograms recorded in such areas do appear approximately as a sequence of primary events which accurately depict the sedimentary layering. Such seismograms could be interpreted in the form in which they were recorded, and as a result they did not require any signal processing. This fortunate situation resulted in the adaption in the 1930s of the reflection seismic method as the premier method for oil exploration. Many of the great oilfields discovered in Texas, Oklahoma, and other parts of the United States during the 1930's and 1940's actually represented such optimal areas, and the seismograms recorded there were made up of a clear-cut series of identifiable primary events neither seriously attenuated by transmission losses nor hopelessly masked by multiples. Of course, a great many other areas produced seismograms that were so confused that they could not be interpreted by any known visual method. Of course, most seismograms fell in between these two extremes, and their interpretation before the advent in the 1950's of digital signal processing required much careful and painstaking work.

The purpose of deconvolution is to take a non-optimal situation and turn it into an optimal one. That is, the purpose of deconvolution is to remove the effects of transmission losses and multiples from the observed seismogram. In order to describe how deconvolution works

Chapter 2. Seismic models

we must further develop our model of the earth as a stack of sedimentary layers.

As we have seen, a seismogram is made up of three types of entities, namely,

(1) the pure primary event which is the reflection coefficient of the given interface,

(2) the product of two-way transmission coefficients of the direct path from the surface to the given interface and back to the surface, and

(3) the myriad of multiples.

In the case of a white (i.e. non-autocorrelated) reflectivity function, entities (2) and (3) cancel each other and we are left with entity (1), the reflection coefficients. However in the general case (i.e. a reflectivity function whose autocorrelation is not a spike), we have all three entities. We will now show that it is possible to simplify this problem.

As we have noted, it is a physical fact that reflection coefficients occurring in sedimentary basins in the earth's crust generally are small in magnitude. In such a case the model for the observed seismogram can be considerably simplified. Specifically, two things happen. They are

(1) the products of the two-way transmission coefficients disappear, and

(2) the myriad of multiple events associated with the sedimentary interfaces are replaced for each interface by a common multiple train, namely, the multiple train associated with the entire layered system.

Admittedly, the above simplification is just an approximation, but it is one that makes clear how the mathematics of deconvolution works. Thus, according to this simplification, the observed seismogram consists of events, each event being made up of a reflection coefficient and a given train of multiples. This train of multiples, as we have said, is the same for each reflection coefficient. But what is this common multiple train? It turns out that this common multiple train is, in fact, the transmission response of the entire layered system. In other words, the transmission response tells us how the whole layered system reverberates. Unfortunately, we cannot put receivers at depth and record the transmission response in a seismic survey.

We have reduced the model for the reflection seismogram to a very simple structure, namely, a series of reflection coefficients to each of which is attached the same (or common) multiple train. We recall that the series of reflection coefficients make up the so-called reflectivity function. Moreover the common multiple train, as we have just stated, is the transmission response of the layered system. The attachment of the same multiple train to each reflection coefficient represents the mathematical process of convolution. Thus in mathematical terms we have the following model for the reflection seismogram:

> Seismogram = (Reflection coefficient series) * (Common multiple train)

or, what is the same thing,

> Reflection response = (Reflectivity function) * (Transmission response).

Here the asterisk indicates the mathematical operation of convolution. This representation of the seismogram is the so-called convolutional model. The important point to remember here is that the more general model of a seismic trace is time-varying. The convolutional model represents an important simplification which holds in the case of small reflection coefficients. In practice we generally assume that the convolutional model only holds over a specified time gate of the seismogram, rather than the entire seismogram. Also we would convolve into the model a source wavelet. In our discussion here we assume that the source wavelet is a spike, so as not to complicate the general ideas of the model with additional conditions.

Reverberations

Seismic waves are mechanical waves that travel through the earth from one location to another. The motion through the rock occurs as one particle of the rock interacts with its neighboring particle. Each moving particle successively transmits its mechanical motion and corresponding energy to its neighbors. This transport of mechanical energy through a medium by particle interaction is what makes a seismic wave a mechanical wave. Electromagnetic waves, which can travel through a vacuum, need no medium for propagation.

As a seismic wave reaches an interface between two rock layers, it undergoes certain characteristic behaviors, such as transmission, reflection and diffraction. The reflection of seismic waves from an interface results in some observable behaviors that have their counterparts in everyday sound waves. In fact, a seismic wave can be roughly compared to a sound wave, not through the air, but through a solid medium. In a large canyon, shortly after shouting, you hear the echo, a less distinct sound that resembles the original shout. This echo results from the reflection of sound off the canyon walls and its ultimate return to your ear. If the canyon wall is more than about seventeen meters away, then the sound wave will take more than 0.1 second to make the round-trip passage. Since the perception of a sound generally endures in memory for only about 0.1 second, there will be a small time delay between the perception of the original sound and the perception of the reflected sound, the echo.

A sound reverberation is perceived in a different way than an echo. A reverberation is heard when the reflected sound wave reaches the ear in less than 0.1 second after the original sound wave. Since the original sound wave is still held in memory, there is no time delay between the perception of the reflected sound wave and the original sound wave. The two sound waves tend to combine as one very prolonged sound wave. Singing in the shower provides a good example. The sound that you hear is the result of the repeated reflection of the sounds between the shower walls. Because the walls are much less than 17 meters apart, these multiple reflections of the sound waves combine to create a prolonged sound, a reverberation.

A reverberation is a result of multiple reflections. A sound wave in an enclosed or semi-enclosed environment will be broken up as it is bounced back and forth among the reflecting surfaces. A sound reverberation is, in effect, a multiplicity of echoes whose speed of repetition is too quick for them to be perceived as separate from one another. In seismology, a reverberation is the multiple bouncing of waves between layers. Short-period seismic reverberations correspond to the sound reverberations. In such a case the successive seismic multiples blend together into a more-or-less steady oscillation. Waves at vertical incidence in a water layer can develop so-called long-period reverberations made up of clear periodic multiple echoes.

Basic to the understanding of the processing of reflection seismic data is the development of a model. A fundamental model is the layered earth model in which the earth consists of horizontal layered strata. First we must look at a single horizontal interface between two layers. A seismic pulse carries energy through the first layer. But when the pulse reaches the interface, where does the energy go? A portion of its energy is transmitted into the second layer (in the form of a transmitted pulse), and a portion of its energy remains in the first layer (in the form of a reflected pulse). Without absorption, the incident energy is divided up into the energy of the reflected pulse and the energy of the transmitted pulse. Since the total energy carried by the incident pulse is divided two ways at the interface, the reflected pulse can have no more energy than the incident pulse. In fact, the energy of the reflected pulse will be less than that to the incident pulse, unless the interface is a perfect reflector so that no energy is transmitted.

Depending upon the impedance contrast of the two strata, the reflected wave may or may not be inverted from that of the incident wave. If the incident wave has a given frequency, then the reflected wave will have the same frequency. Comparisons can also be made between the characteristics of the transmitted wave and those of the incident pulse. First, the transmitted wave is never inverted. In fact inversion can only occur for the reflected wave, if it occurs at all. Second, when waves cross interfaces the frequency of the transmitted wave is the same as the frequency of the incident wave. The reason is that the vibration of the last particle in the incident layer creates the vibration of the first particle on the opposite side of the interface. These two particles are adjoined in such a manner that the frequency at which one particle vibrates is equal to the frequency at which the other particle vibrates. This handshake principle is the reason why the frequency of the incident pulse and the transmitted pulse is the same.

Layered earth model

The most pronounced variations in the earth layering are along the vertical scale, a one-dimensional vertical model. The foremost 1D model is the stratified or layered-earth model. This model uses discrete horizontal layers to represent the earth. It gives a useful approximation for the propagation of vertically-traveling waves

The internal structure of an inhomogeneous earth can be characterized by its impedance function. For the layered-earth model, it is assumed that the impedance is a function of the vertical dimension (z axis) only. It is convenient to measure on the z axis, not distance, but traveltime. In this way, distance and time are measured in terms of the same unit. This unit, known as the wave-second, represents the distance that a material wave propagates in one second. Consider a layer in which the seismic wave velocity is 4000 m/s, or, in other words, 4 m/ms. For such a layer, one millisecond would represent a distance of 4 m. The use of this space-time convention means that the same scale can be used for both the vertical z axis of the earth and the time t axis of the various signals.

The layered-earth model represents a discrete approximation to a continuous inhomogeneous medium. In this model, the earth is mathematically sliced into many thin horizontal layers normal to the vertical z direction. This theoretical division of the earth into thin layers produces a stratified medium characterized by the interfaces between the layers. An arbitrary time unit is chosen, and the thickness of each layer is chosen as one-half of the time unit. The reason for this choice of thickness is so that the two-way travel time (downward time plus upward time) in each layer is equal to one unit. In the mathematical construction of the layered-earth model, it is assumed that the impedance is constant within each layer. However, the impedance can change from layer to layer. As is well known, an impedance contrast gives rise to a reflection coefficient. Each interface separating two adjacent layers with different impedances has a non-zero reflection coefficient. The greater the impedance contrast between the two adjacent layers, the greater is the magnitude of the reflection coefficient. The wave motion is digitized so that a signal becomes a discrete sequence of pulses. The pulse amplitudes are measured in terms of particle velocity in the case of a geophone and in terms of pressure in the case of a hydrophone. It is assumed that there is no dissipation of kinetic energy into heat. Thus all the source energy imparted into the body can be accounted for, over time, in terms of the resulting elastic wave motion.

Each reflection coefficient is a local reflection coefficient, or a so-called Fresnel reflection coefficient. Such a reflection coefficient characterizes the interface separating two adjacent strata in isolation from all the other strata. The Fresnel reflection coefficient depends only on the impedances of the two strata in question, and not on the impedances of the remaining strata. A pulse normally incident on such an isolated interface is divided into a reflected pulse and a transmitted pulse. Energy is conserved. As a result, the energy incident on an interface is equal to the sum of the energy transmitted through the interface and the energy reflected from the interface. As a consequence, the magnitude of the Fresnel reflection coefficient must less than one. The magnitude of the reflection coefficient does not depend upon the direction in which the pulse travels through the interface, but the sign of the reflection coefficient does. For a given interface, the reflection coefficient for a upgoing pulse is the negative of the reflection coefficient for a downgoing pulse.

We consider plane-wave motion traveling normal to the interfaces, that is, waves traveling in the vertical z direction. In this model, the sequence of Fresnel reflection coefficients associated with the entirety of the interfaces represents the internal structure of the earth. Thus what we call the internal structure is the sequence of Fresnel reflection coefficients for a downgoing pulse. Mathematically, the internal structure is represented by a sequence of numbers each less than one in magnitude. Let N be the number of interfaces which mathematically slice the inaccessible earth. Each interface k where $k = (1, 2, \cdots, N)$ has a reflection coefficient c_k for a downgoing incident wave.

These reflection coefficients are either for particle velocity or pressure, as the case may be. The sequence of reflection coefficients

$$c = (c_1, c_2, \cdots, c_N)$$

is called the reflectivity function or simply the reflectivity. The reflectivity, once it is determined, can be used to compute the required impedance function of the body in a straight-forward manner. The reflectivity is the unknown quantity in the remote sensing problem faced in seismic prospecting. The reflection coefficient for a downgoing wave incident on interface k is

$$c_k$$

The reflection coefficient c_k' for an upgoing wave incident on interface k is

$$c_k' = -c_k$$

The transmission coefficient τ_i for downward transmission through interface k is

$$\tau_k = 1 + c_k$$

The transmission coefficient τ_i' for upward transmission through interface k is

$$\tau_k' = 1 + c_k'$$

The one-way transmission factor through n interfaces (from top to bottom) is

$$\sigma_N = (1 + c_1)(1 + c_2) \cdots (1 + c_N)$$

The one-way transmission factor through n interfaces (from bottom to top) is

$$\sigma_N' = (1 - c_1)(1 - c_2) \cdots (1 - c_N)$$

The two-way transmission factor

$$\sigma_N \sigma_N' = (1 - c_1^2)(1 - c_2^2) \cdots (1 - c_N^2)$$

lies between zero and one.

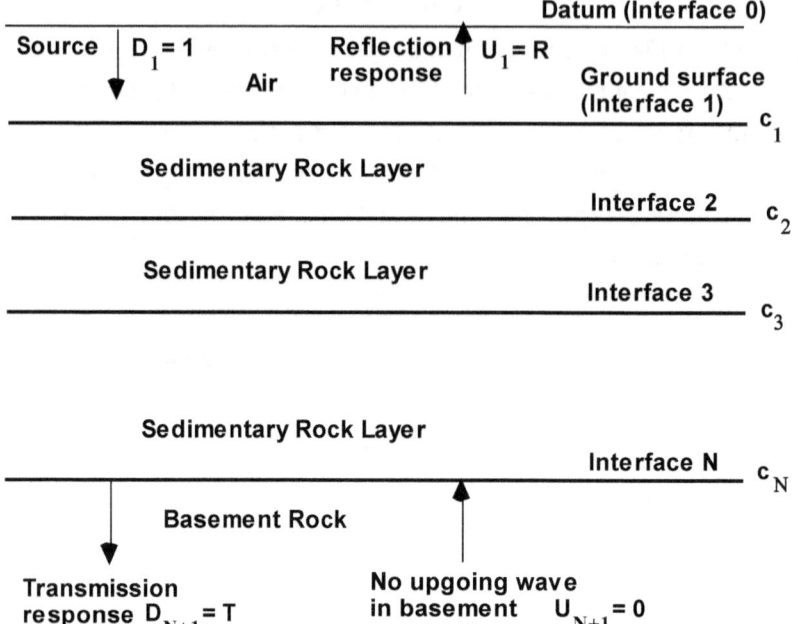

Fig. 6. A layered system with N interfaces

In the stratified model the boundary layers are the air (on the top) and the basement rock (on the bottom). See Fig. 1. All the plane-wave motion is sandwiched between the air and the basement rock. In the model no internal loss of energy by absorption within the layers is allowed. Because energy degradation effects are neglected, the layered system represents a lossless system in which energy leaves the system only by net transmission downward into the basement or net reflection upward into the air. Thus the input energy from a downgoing unit spike at the surface is divided between the wave transmitted by the layered system into the basement and the wave reflected by the layered system into the air. If the input is a unit downgoing spike (i.e., a unit impulse), then the transmitted wave is called the transmission impulse response, and the reflected wave is called the reflection impulse response.

For convenience, the source and receiver are mathematically placed on a datum, called interface 0, which is one-half time unit above the surface (which is interface 1). The reflection coefficient of the datum is taken to be zero. Consider the case for which the stratified system has one input or source (a downgoing plane wave on the datum at normal incidence) and two outputs (a downgoing plane wave emitted from the bottom interface and an upgoing plane wave received on the datum). The source, activated at time zero, gives rise to a downgoing plane wave that represents the input into the body. This input is called the source wavelet or source signature. In conventional seismic exploration, the observed data are the resulting signals reflected upward from depth and recorded at or close to the surface. The reflection coefficients are determined by the impedance contrasts between the rock layers within the earth. The values of the reflection coefficients represent the desired information we wish to find by analysis of the received reflected data.

Let $D_k(Z)$ and $U_k(Z)$ be respectively the transfer functions of the downgoing wave and the upgoing wave at a point just below interface $k-1$. The Lorentz transform between two adjacent layers can be written in matrix form as

$$\begin{pmatrix} D_{k+1}(Z) \\ U_{k+1}(Z) \end{pmatrix} = \frac{Z^{-1/2}}{\tau'_k} \begin{pmatrix} Z & -c_k \\ -c_k Z & 1 \end{pmatrix} \begin{pmatrix} D_k(Z) \\ U_k(Z) \end{pmatrix}$$

Define the fundamental polynomials $P_k(Z)$ and $Q_k(Z)$ and their delayed-reverses

$$P_k^R(Z) = Z^k \, P_k(Z^{-1})$$
$$Q_k^R(Z) = Z^k \, Q_k(Z^{-1})$$

by the recursion

$$\begin{pmatrix} P_k^R(Z) & Q_k^R(Z) \\ Q_k(Z) & P_k(Z) \end{pmatrix} = \begin{pmatrix} Z & -c_k \\ -c_k Z & 1 \end{pmatrix} \begin{pmatrix} Z & -c_{k-1} \\ -c_{k-1} Z & 1 \end{pmatrix} \cdots \begin{pmatrix} Z & -c_1 \\ -c_1 Z & 1 \end{pmatrix}$$

which gives

$$\begin{pmatrix} P_k^R(Z) & Q_k^R(Z) \\ Q_k(Z) & P_k(Z) \end{pmatrix} = \begin{pmatrix} Z & -c_k \\ -c_k Z & 1 \end{pmatrix} \begin{pmatrix} P_{k-1}^R(Z) & Q_{k-1}^R(Z) \\ Q_{k-1}(Z) & P_{k-1}(Z) \end{pmatrix}$$

The recursion can be written as

$$P_k(Z) = P_{k-1}(Z) - c_k Z \, Q_{k-1}^R(Z)$$
$$Q_k(Z) = Q_{k-1}(Z) - c_k Z \, P_{k-1}^R(Z)$$

The sequence of fundamental polynomials $P_k(Z)$ and $Q_k(Z)$ for $k = 1, 2, \cdots, N$ characterize the stratified system. Although $P_k(Z)$ is actually a polynomial of degree $k-1$, it is treated as if it were a polynomial of degree k with last coefficient 0. The first coefficient of $Q_k(Z)$ is zero.

The Lorentz transform in the form that gives the relationship between the waves in layer $N+1$ to the waves in layer 1 is

$$\begin{pmatrix} D_{N+1}(Z) \\ U_{N+1}(Z) \end{pmatrix} = \frac{Z^{-N/2}}{\sigma'_N} \begin{pmatrix} P_N^R(Z) & Q_N^R(Z) \\ Q_N(Z) & P_N(Z) \end{pmatrix} \begin{pmatrix} D_1(Z) \\ U_1(Z) \end{pmatrix}$$

Because energy absorption effects are neglected, the layered system represents a lossless system in which energy leaves the system only by net transmission downward into the basement rock or net reflection upward into the air. If the input is a unit downgoing spike (i.e., an unit impulse), then the transmitted wave is called the transmission impulse response, and the reflected wave is called the reflection impulse response. The upgoing reflected wave train recorded by the receiver at the datum is denoted by

$$r = (r_1, r_2, r_3, \cdots)$$

and is called the reflection impulse response. Note that the first non-zero coefficient r_1 occurs at time index 1, because it takes one time unit for the pulse to travel from source on the datum to the surface of the earth and back to the receiver on the datum. The Z-transform of the reflection impulse response is

$$R(Z) = r_1 Z + r_2 Z^2 + r_3 Z^3 + \cdots = U_1(Z)$$

The downgoing wave train transmitted into the basement (just below interface$N$) is denoted by

$$t = (t_{(N/2)}, t_{(N/2)+1}, t_{(N/2)+2}, \cdots)$$

and is called the transmission impulse response. (The usage of the letter t for transmission should not be confused with the familiar usage of t for time.) The Z-transform of the transmission impulse response is

$$T(Z) = t_{(N/2)} Z^{N/2} + t_{(N/2)+1} Z^{(N/2)+1} + t_{(N/2)+2} Z^{(N/2)+2} + \cdots U_{N+1}(Z)$$

The first nonzero amplitude of the transmission response is at $N/2$, because it takes $N/2$ time units for the one-way trip through the N layers.

Consider the forward problem. The initial conditions are:

(1) the source is a downgoing unit spike (an impulsive unit source) initiated at time 0 on the datum, so $D_1(Z)$, and

(2) there is no upgoing wave motion in the basement, so $U_{N+1}(Z) = 0$.

It is required to find the transfer function

$$T(Z) = U_{N+1}(Z)$$

of the transmission impulse response and the transfer function

$$R(Z) = U_1(Z)$$

of the reflection impulse response. Application of the Lorentz transform gives

$$\begin{pmatrix} T(Z) \\ 0 \end{pmatrix} = \frac{Z^{-N/2}}{\sigma'_N} \begin{pmatrix} P_N^R(Z) & Q_N^R(Z) \\ Q_N(Z) & P_N(Z) \end{pmatrix} \begin{pmatrix} 1 \\ R(Z) \end{pmatrix}$$

The solution of these equations gives the transfer function of the transmission impulse response as

$$T(Z) = \frac{\sigma_N Z^{N/2}}{P_N(Z)}$$

and the transfer function of the reflection impulse response as

$$R(Z) = \frac{-Q_N(Z)}{P_N(Z)}$$

This expression for the reflection impulse response is called the nonlinear model. It is seen that the reflectivity function determines the reflection response (i.e., seismic trace) and the transmission response (i.e., the waveform which is transmitted into the basement and never returned to the layered system). Because physically the transmission impulse response is a causal transient time function, it follows that $P_N(Z)$ is minimum delay. However, the minimum-delay property of $P_N(Z)$ can also be established mathematically by using the fact that each reflection coefficient is less than one in magnitude.

Let us define the multiple wave train

$$m = (m_0, m_1, m_2, \cdots)$$

by means of the Z-transform

$$M(Z) = \sum_{k=0}^{\infty} m_k Z^k = \frac{1}{P_N(Z)}$$

Because the polynomial $P_N(Z)$ is minimum-delay, it follows that $M(Z)$ is also minimum-delay. The reflection impulse response becomes

$$R(Z) = \frac{-Q_N(Z)}{P_N(Z)} = -Q_N(Z)M(Z)$$

The transmission impulse response becomes

$$T(Z) = \frac{\sigma_N Z^{N/2}}{P_N(Z)} = \sigma_N Z^{N/2} M(Z)$$

The above equation states: The transmission impulse response is made up of the direct pulse σ_N from the shot to (and through) interface N and the pulses of the attached multiple reflections.

The direct pulse has amplitude σ_N (i.e., the downward transmission factor) and is delayed by the traveltime $0.5N$ downward through the N layers (as given by the delay factor $Z^{N/2}$). The attached multiple reflections are generated by the feedback system given by $1/P_N(Z)$.

Einstein addition

The behavior of a wave at an interface can summarized as follows: the energy of the reflected pulse is less than (or at most equal to) the energy of the incident pulse, the reflected pulse may become inverted but not the transmitted pulse, and the frequency is not altered by crossing the interface. In the case of a vertically incident wave on a horizontal interface, the behavior may be put in terms of a single number, the reflection coefficient c_k of interface k. This coefficient must have a value that lies between -1 and 1.

Let the wave incident on interface k be a downgoing pulse of amplitude A. The resulting reflected wave is an upgoing pulse of amplitude $c_k A$ and the resulting transmitted wave is an downgoing pulse of amplitude $\tau_k A$. Let the wave incident on interface k be a upgoing pulse of amplitude A. The resulting reflected wave is an downgoing pulse of amplitude $c'_k A$ and the resulting transmitted wave is an upgoing pulse of amplitude $\tau'_k A$.

Chapter 2. Seismic models

Now let us consider the case of one interface (interface 1) with reflection coefficient c_1. The two-way travel time for the layer between any two consecutive interfaces is 1 time unit. Thus the one-way travel time for the layer between the two consecutive interfaces is 1/2 time unit. We will measure all wave motion at a point just below the interface. For this reason, the datum is an hypothetical (non-existent) interface located 1/2 time unit above the surface (interface 1). Let the input be a unit spike. It will appear just below the datum. The reflected pulse c_1 will be received just below the datum on time unit later. The reason is that it takes one time unit for the pulse to travel from the datum to the surface and then back to the datum.

We assume normal incidence. Let a downgoing incident unit pulse strike the interface 1 at time 1/2. The reflection from the interface 1 will be recorded at the datum at time 1. The mathematical symbol Z represents the time delay 1. The delay in the one way travel time in going from one interface to the next is represented by $Z^{1/2}$. The transmission and reflection responses can be written as

$$T_1(Z) = \tau_1 Z^{1/2} \quad \text{and} \quad R_1(Z) = c_1 Z$$

We will write these expressions as

$$T_1(Z) = \frac{\tau_1 Z^{1/2}}{1} \quad \text{and} \quad R_1(Z) = -\frac{-c_1 Z}{1}$$

Define the first-order polynomials as

$$P_1(Z) = 1 \quad \text{and} \quad Q_1(Z) = -c_1 Z$$

Thus

$$T_1(Z) = \frac{\tau_1 Z^{1/2}}{P_1(Z)} \quad \text{and} \quad Q_1(Z) = -\frac{Q_1(Z)}{P_1(Z)}$$

Now we must consider the case of two interfaces, the top (interface 1) with reflection coefficient c_1 and the bottom (interface 2) with reflection coefficient c_2. As before, we assume normal incidence. Let a downgoing incident unit pulse strike the interface 1 at time 1/2. The reflection from the interface 1 will be recorded at the datum at time 1. The primary reflection from interface 2 will be recorded at the datum at time 2. The direct transmission starts at the datum, goes through interface 1, and the goes through interface 2. The direct transmission is recorded just below interface 2 at time 1.

For a downward unit spike incident on the interface, the transmission and reflection responses are

$$T_2(Z) = \frac{\tau_1 \tau_2 Z}{1 + c_1 c_2 Z} \quad \text{and} \quad R_2(Z) = \frac{c_1 Z + c_2 Z^2}{1 + c_1 c_2 Z}$$

Define second-order polynomials as

$$P_2(Z) = 1 + c_1 c_2 Z \quad \text{and} \quad Q_2(Z) = -c_1 Z - c_2 Z^2$$

Thus

$$T_2(Z) = \frac{\tau_1 \tau_2 Z}{P_2(Z)} \quad \text{and} \quad R_2(Z) = -\frac{Q_2(Z)}{P_2(Z)}$$

If the second interface were absent, then the complete reflection response would be the primary reflection represented by

$$c_1 Z$$

If the first interface were absent (all else remaining the same), then the complete reflection response would the nominal primary reflection represented by

$$c_2 Z^2$$

For the moment neglect all other considerations. Then the refection response for the layer would be the sum given by Newton addition; namely, the sum of the two reflections (without transmission effects or multiples)

$$c_1 Z + c_2 Z^2$$

This formula is routinely used successfully on a routine basis in studies of synthetic seismograms. However, for accuracy Einstein addition should be used.

First of all let us factor out the Z in the above equation. We obtain

$$Z(c_1 + c_2 Z)$$

The factored-out Z is an artifact due to our choice of the datum as being 1/2 time unit above the surface of the earth. This arbitrary choice was made for the mathematical convenience.

Chapter 2. Seismic models

We will use the Einstein addition formula on the expression within the parentheses, namely the addition of the two terms c_1 and $c_2 Z$. The result of Einstein addition is a quotient. The numerator is the sum of the two terms; namely,

$$c_1 + c_2 Z$$

The denominator is sum of 1 plus the product of the two terms, namely,

$$1 + c_1 c_2 Z$$

Thus the Einstein addition is

$$\frac{c_1 + c_2 Z}{1 + c_1 c_2 Z}$$

We now put back the factored-out Z. The result is the refection response for the layered system; namely, the Einstein formula

$$\frac{c_1 Z + c_2 Z^2}{1 + c_1 c_2 Z}$$

We conclude with an interpretation of the Einstein formula. A little algebra converts the Einstein formula to the form

$$\frac{c_1 Z + c_2 Z^2}{1 + c_1 c_2 Z} = c_1 Z + c_2 \frac{(1 - c_1^2) Z^2}{1 + c_1 c_2 \, Z}$$

The initial c_1 represents the primary reflection from the top interface. The factor $(1 - c_1^2)$ represents the transmission loss suffered by the reflection upon crossing interface 1. To this reflection are attached the reverberations as given by the Einstein denominator $1 + c_1 c_2 \, Z$. If we expand this denominator in a geometric series the Einstein formula becomes

$$c_1 Z + c_2 (1 - c_1^2) Z^2 - c_2 (1 - c_1^2) c_1 c_2 Z^3 + c_2 (1 - c_1^2) \, c_1^2 \, c_2^2 \, Z^4 - \cdots$$

The first term is the first primary reflection, the second term is the second primary reflection, the third term is the first multiple reflection of the second primary, the fourth term is the second multiple reflection of the second primary, and so on. All of these multiples make up the reverberation attached to the second primary.

Mathematical examples

Let $c_1, c_2, c_3, \cdots, c_n$ be the set of Fresnel reflection coefficients. The set is called the reflectivity series. Let $\gamma_1, \gamma_2, \gamma_3, \cdots, \gamma_{n-1}$ be the set of autocorrelation of the reflectivity series.

The only source of energy is the initial downgoing unit spike which gives rise to a net downgoing flow of energy. As there are no sources or sinks in any layer, the net downgoing energy spectrum must be the same in each layer. As we have seen,

$$R_2(Z) = -\frac{Q_2(Z)}{P_2(Z)} \quad \text{and} \quad T_2(Z) = \frac{(1+c_1)(1+c_2)Z}{P_2(Z)}$$

From two sections above, we know

$$R_N(Z) = \frac{-Q_N(Z)}{P_N(Z)} \quad \text{and} \quad T_N(Z) = \frac{\sigma_N Z^{N/2}}{P_N(Z)}$$

For example, for $N = 4$ interfaces, we have

$$P_4(Z) = 1 + (c_1c_2 + c_2c_3 + c_3c_4)Z + (c_1c_3 + c_2c_4 + c_1c_2\,c_3c_4)Z^2 + c_1c_4Z^3$$

For small reflection coefficients, the term $c_1c_2\,c_3c_4$ is of higher order and can be neglected. In case the reflection coefficients are small in magnitude, then $P_4(Z)$ is approximately

$$P_4(Z) = 1 + (c_1c_2 + c_2c_3 + c_3c_4)Z + (c_1c_3 + c_2c_4)Z^2 + c_1c_4Z^3$$

We note that

$$c_1c_2 + c_2c_3 + c_3c_4 = \gamma_1$$
$$c_1c_3 + c_2c_4 = \gamma_2 Z^2$$
$$c_1c_4 = \gamma_3$$

where $\gamma_1, \gamma_2, \gamma_3$ are autocorrelation coefficients of the reflection coefficient sequence c_1, c_2, c_3, c_4. As a result

$$P_4(Z) \approx 1 + \gamma_1 Z + \gamma_2 Z^2 + \gamma_3 Z^3$$

More generally, for small reflection coefficients we have approximately

$$P_N(Z) = 1 + \gamma_1 Z + \gamma_2 Z^2 + \cdots + \gamma_{N-1} Z^{N-1}$$
$$Q_N(Z) = -c_1 Z - c_2 Z^2 - \cdots - c_N Z^N$$

Thus the transmission response becomes approximately

Chapter 2. Seismic models

$$T(Z) = \frac{\sigma_N Z^{N/2}}{1 + \gamma_1 Z + \gamma_2 Z^2 + \cdots + \gamma_{N-1} Z^{N-1}}$$

which is a feedback system. The reflection response becomes approximately

$$R(Z) = \frac{c_1 Z + c_2 Z^2 + \cdots + c_k Z^N}{1 + \gamma_1 Z + \gamma_2 Z^2 + \cdots + \gamma_{k-1} Z^{N-1}}$$

If we define $C(Z)$ as

$$C(Z) = c_1 Z + c_2 Z^2 + \cdots + c_N Z^N$$

and $W(Z)$ as

$$W(Z) = \frac{1}{1 + \gamma_1 Z + \gamma_2 Z^2 + \cdots + \gamma_{N-1} Z^{N-1}}$$

then

$$R(Z) = C(Z)W(Z)$$

What does this equation tell us? It contains three components: namely,

(1) $R(Z)$, which represents the upgoing reflection seismogram recorded on the surface of the earth resulting from a downgoing unit spike introduce at the surface of the earth,

(2) $C(Z)$, which represents the reflection coefficients of the horizontal subsurface interfaces,

(3) $W(Z)$, which represents the common multiple train.

In conclusion, the equation $R(Z) = C(Z)W(Z)$ represents the generating function of the reflection seismogram as the product of the generating function of the reflectivity and the generating function of the common wavelet.

Numerical example

Suppose the reflectivity is given by the vector

$$c = \begin{pmatrix} 0.1 \\ -0.5 \\ 0.25 \\ -0.02 \\ 0 \\ 0 \end{pmatrix}$$

Define the reflectivity matrix C as

$$C = \begin{pmatrix} 0.1 & 0 & 0 & 0 & 0 & 0 \\ -0.5 & 0.1 & 0 & 0 & 0 & 0 \\ 0.25 & -0.5 & 0.1 & 0 & 0 & 0 \\ -0.2 & 0.25 & -0.5 & 0.1 & 0 & 0 \\ 0 & -0.2 & 0.25 & -0.5 & 0.1 & 0 \\ 0 & 0 & -0.2 & 0.25 & -0.5 & 0.1 \end{pmatrix}$$

The autocorrelation matrix is

$$C^T C = \begin{pmatrix} 0.3625 & -0.225 & 0.125 & -0.02 & 0 & 0 \\ -0.225 & 0.3625 & -0.225 & 0.125 & -0.02 & 0 \\ 0.125 & -0.225 & 0.3625 & -0.225 & 0.125 & -0.02 \\ -0.02 & 0.125 & -0.225 & 0.3225 & -0.175 & 0.025 \\ 0 & -0.02 & 0.125 & -0.175 & 0.26 & -0.05 \\ 0 & 0 & -0.02 & 0.025 & -0.05 & 0.01 \end{pmatrix}$$

We modify this matrix by setting the elements on the main diagonal equal to one. We hereby obtain

$$\Gamma = \begin{pmatrix} 1 & -0.225 & 0.125 & -0.02 & 0 & 0 \\ -0.225 & 1 & -0.225 & 0.125 & -0.02 & 0 \\ 0.125 & -0.225 & 1 & -0.225 & 0.125 & -0.02 \\ -0.02 & 0.125 & -0.225 & 1 & -0.175 & 0.025 \\ 0 & -0.02 & 0.125 & -0.175 & 1 & -0.05 \\ 0 & 0 & -0.02 & 0.025 & -0.05 & 1 \end{pmatrix}$$

The elements in the first column give the vector

$$\gamma = \begin{pmatrix} 1 \\ \gamma_1 \\ \gamma_2 \\ \gamma_3 \\ \gamma_4 \\ \gamma_1 \end{pmatrix} = \begin{pmatrix} 1 \\ -0.225 \\ 0.125 \\ -0.02 \\ 0 \\ 0 \end{pmatrix}$$

The reflection response is

$$R(Z) = C(Z)W(Z)$$

Define

$$W = \Gamma^{-1} = \begin{pmatrix} 1.06 & 0.22 & -0.09 & -0.02 & 0.01 & -0 \\ 0.22 & 1.11 & 0.2 & -0.09 & -0.02 & 0.01 \\ -0.09 & 0.2 & 1.12 & 0.21 & -0.1 & 0.01 \\ -0.02 & -0.09 & 0.21 & 1.09 & 0.16 & -0.01 \\ 0.01 & -0.02 & -0.1 & 0.16 & 1.04 & 0.05 \\ -0 & 0.01 & 0.01 & -0.01 & 0.05 & 1 \end{pmatrix}$$

The elements in the first column give the vector

Chapter 2. Seismic models

$$\mathbf{w} = \begin{pmatrix} w_0 \\ w_1 \\ w_2 \\ w_3 \\ w_4 \\ w_5 \end{pmatrix} = \begin{pmatrix} 1.06 \\ 0.22 \\ -0.09 \\ -0.02 \\ 0.01 \\ -0 \end{pmatrix}$$

Next compute the reflection seismogram

$$R(Z) = W(Z)C(Z)$$

by means of the matrix equation

$$\mathbf{r} = \mathbf{Wc} = \begin{pmatrix} -0.03 \\ -0.48 \\ 0.16 \\ 0.07 \\ -0.02 \\ 0 \end{pmatrix}$$

The reflection seismogram

$$r = (-0.03, -0.48, 0.16, -0.02, 0, 0)$$

is the convolution of the reflectivity

$$c = (0.1, -0.5, 0.25, -0.02, 0, 0)$$

and the wavelet

$$w = (1.06, 0.22, -0.09, -0.02, 0.01, 0)$$

The reflectivity can be recovered from the reflection seismogram by means of deconvolution. We compute the reflectivity

$$C(Z) = C(Z)W(Z)$$

by means of the matrix equation

$$\mathbf{c} = \mathbf{W}^{-1}\mathbf{r} = \mathbf{\Gamma r} = \begin{pmatrix} -0.03 \\ -0.48 \\ 0.16 \\ 0.07 \\ -0.02 \\ 0 \end{pmatrix}$$

The reflectivity

$$c = (0.1, -0.5, 0.25, -0.02, 0, 0)$$

is the convolution of the reflection seismogram

$$r = (-0.03, -0.48, 0.16, -0.02, 0, 0)$$

and the inverse wavelet

$$\gamma = (1, -0.225, 0.125, -0.02, 0, 0)$$

Common multiple train

Let us now examine the common multiple train; i.e., the common wavelet that is attached to each primary reflection. Define the wavelet by the generating function $W(Z)$. Then the generating function of the reflection response is

$$R(Z) = c_1 Z\, W(Z) + c_2 Z^2\, W(Z) + \cdots + c_N Z^N\, W(Z)$$

In the time domain, this equation becomes the convolution

$$r_t = c_1 w_{t-1} + c_2 w_{t-2} + \cdots + c_N w_{t-N} = \sum_{k=1}^{N} c_k w_{t-k}$$

By definition, the common multiple train is the result of the reverberations occurring within the entire layered system. If we examine the types of reverberations which can occur, we can see that each of the reverberations results from a negative feedback loop.

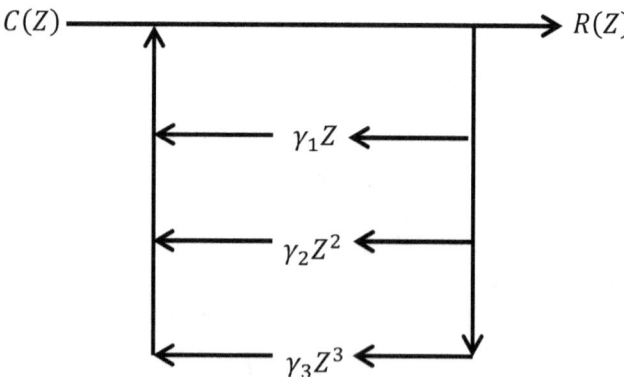

Fig. 7. The convolutional model

The net result is that all the reverberations taken together can be described by a negative feedback system with the autocorrelation function (for positive lags) of the reflectivity within the feedback box. See Figs. 7 and 8.

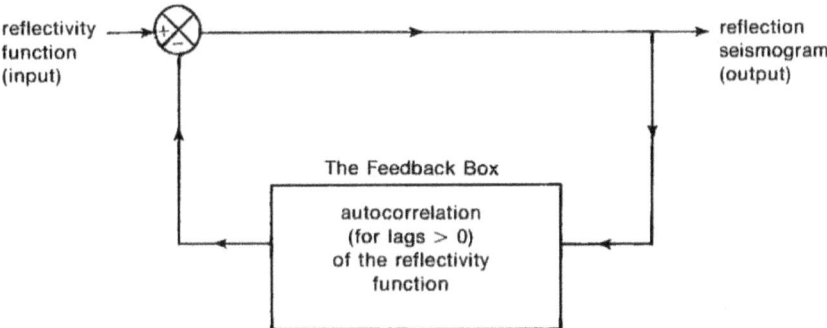

Fig. 8. The convolutional model

Deconvolution

The problem of deconvolution in seismic data processing can be stated simply as follows. Given the reflection seismogram (which is observed at the surface of the earth), find the reflectivity function (which gives stratigraphic information as a function of depth). In order to find a solution to this problem we will make use of the convolutional model which we assume holds over a given time gate of the seismogram.

As we have seen, the convolutional model is a pure feedback model. A pure feedback system is necessarily a minimum-delay system. More precisely, the model

$$R(Z) = C(Z)W(Z)$$

says that

Seismograph = (Reflectivity function) * (Common multiple train)

We can assert that the common multiple train is a minimum-delay wavelet. Thus one aspect of the seismic convolutional model is that it is a minimum-delay model.

Let us now summarize the results we have obtained up to this point. The convolutional model states that the reflection seismogram is the convolution of the reflectivity function with the common multiple train. As we know the common multiple train is a waveform made up of the reverberations from all the different combinations of the layers. As a result the system whose impulse response is the common multiple train is a negative-feedback system. The feedback box of this direct system is made up of the autocorrelation (for positive lags) of the reflectivity function. Such a feedback system is necessarily minimum delay.

For each minimum-delay system, there is a causal inverse system. In the case of a pure negative feedback system, the causal inverse system is a pure feed-forward system. In our case, the inverse system can be found by inspection. The result is

$$C(Z) = \frac{R(Z)}{W(Z)}$$

which is

$$R(Z) = \left(1 + \gamma_1 Z + \gamma_2 Z^2 + \cdots + \gamma_{k-1} Z^{k-1}\right) R(Z)$$

Specifically, the inverse system is the feed-forward system with the feed-forward box consisting of the autocorrelation (for lags greater than zero) of the reflectivity function.

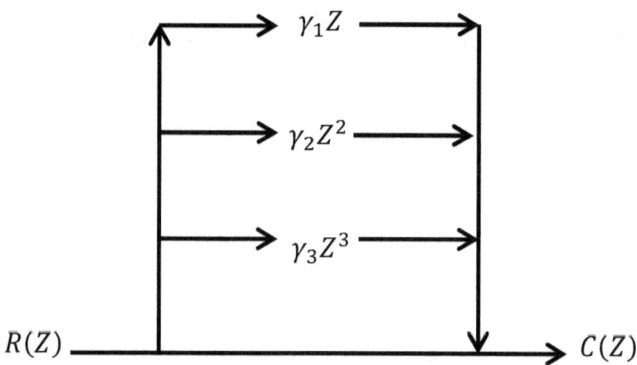

Fig. 9. The deconvolutional model

The net result is that all the reverberations can be eliminated by a feed-forward system with the autocorrelation function (for positive lags) of the reflectivity within the feed-forward box.

See Figs. 9 and 10. The feed-forward box of the inverse system is the same as the feedback box of the direct system. Thus the impulse response of this inverse system is made up of 1 (corresponding to the straight path) together with the positive-lag autocorrelation of the reflectivity function (corresponding to the feed-forward box path). This impulse response represents the operator with which we convolve the reflection seismogram in order to obtain the reflectivity function. That is, the operator converts the observed reflection seismogram to the desired reflectivity function, and hence is the required deconvolution operator.

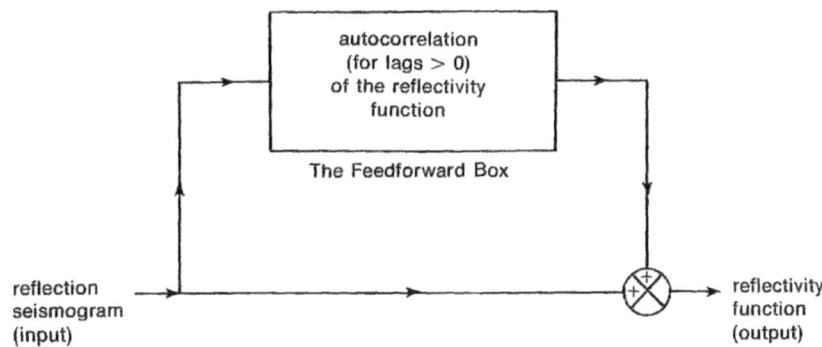

Fig. 10. The deconvolutional model

In brief, the convolutional model states that the reflection seismogram is the convolution of the reflectivity function with the common multiple train. The common multiple train can be regarded as the impulse response of the direct system representing the action of the earth's sedimentary layers. The deconvolution operator is the impulse response of the corresponding inverse system. Thus the deconvolution operator removes the effects of the common multiple train, and thus yields the reflectivity function. Thus we have the direct (or physical) system

   Reflectivity function ⟩ Common multiple train →Reflection seismogram

and the inverse (or data-processing) system

   Reflection seismogram → Deconvolution operator → Reflectivity function

The known (or observational) information is the reflection seismogram, and the desired information is the reflectivity function. We know theoretically that the deconvolution operator is made up of unity followed by the positive-lag values of the autocorrelation of the reflectivity function. Because we do not know the reflectivity function, we must find some way to estimate the deconvolution operator from the known data (i.e. from the reflection seismogram). In order to find a method, we must first introduce the random reflection coefficient model.

As we discussed earlier, many great oilfields in Texas and Oklahoma discovered in the early days of seismic prospecting were in areas which produced textbook-type seismograms. These seismograms as recorded in the field showed beautiful primary reflections which accurately represented the sedimentary structure. The reason is that, in these particular areas, the sedimentary layers give rise to a sequence of reflection coefficients (i.e. the reflectivity function) which is of a white-noise structure. Due to this randomness in the sedimentary column, the multiples all interact with each other and the net effect is that the multiples cancel each other out, except at the times of the primaries, where they build up and actually enhance the primary reflections.

When there are one or more strong reflecting layers in the sedimentary column, then the multiples from these layers build up and mask the primary events. For example, in marine exploration the water layer represents a non-attenuating medium bounded by two strong reflecting interfaces and hence represents an energy trap. A seismic pulse generated in this energy trap will be successively reflected between the two interfaces. Consequently, reflections from deep horizons below the water layer will be obscured by the water reverberations. As another example, a limestone layer at depth with strong reflecting interfaces can also produce multiple reflections which interfere with primary information on the seismogram.

Despite the presence of strong reflecting interfaces interspersed in the geologic column, there remain significant sections where the interfaces are characterized by reflections coefficients that are small and random. Hence on the corresponding sections of the reflection seismogram, the reflectivity function may be considered to be white. Thus by carefully selecting time gates on a reflection seismogram, we are able to pick out sections where we may assume the reflectivity function in a white-noise function. We recall that the convolutional model states that the reflection seismogram is the output of a minimum-delay system. Its impulse response is the multiple-train of the entire sedimentary section and its input is the reflectivity function. By proper selection of the time gate, we see that the input may be considered to be a white-noise series.

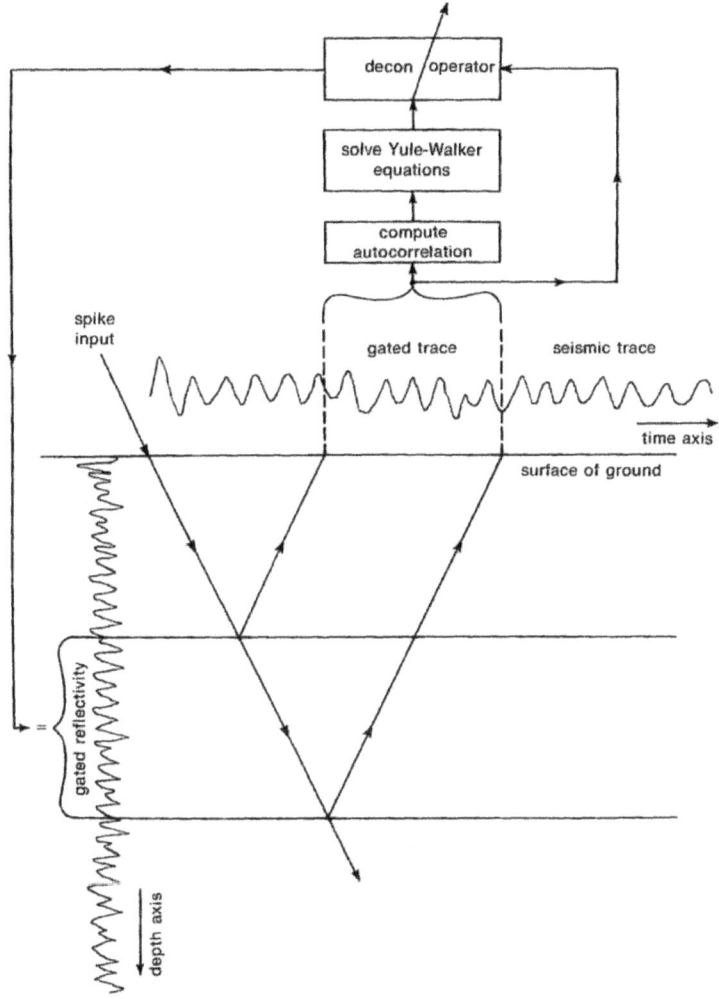

Fig. 11. Predictive deconvolution of a seismic trace. Several gates are used on each trace to obtain a time-varying effect.

See Fig. 11. We have specialized the convolutional model so that it can form the basis for the method to determine the required deconvolution operator. The specialization states that

(1) the earth acts as a minimum-delay system in producing the train of multiple events that appear on the reflection seismogram

(2) the reflectivity function over a selected section of the sedimentary column is a white-noise function

Thus this seismic model differs from any arbitrary convolutional model in that the seismic model is a minimum-delay system with a white-noise input. Because of these special features, the seismic model can be used as a basis for determining the deconvolution operator. In brief, the seismic model is a minimum-delay random-reflection-coefficient convolutional model.

The seismic convolutional model has two characteristic features within the time-gate of interest, namely,

(1) the statistical feature that the primary events are due to a reflectivity function (i.e. series of reflection coefficients) given within the time gate by a random white-noise series, and

(2) the deterministic feature that the multiple wave trains attached to *the primary events have the same minimum-delay wavelet shape.* (Of course, this common multiple train is due to the entire sedimentary section, i.e. to reflection coefficients both within and outside the time-gate.)

The observational data are in the form of the observed seismic trace recorded at the surface of the ground. Let us now discuss the computational procedure used to determine the deconvolution operator (steps (1) and (2) below) and then to carry out the deconvolution (step (3) below).

(1) The first step is to compute the autocorrelation function of that portion of the seismic trace within the specified time-gate

(2) The second step is to compute the coefficients of the prediction error operator. The prediction error operator is in fact the autoregressive operator corresponding to that autocorrelation. This calculation involves solving a set of simultaneous equations called the Yule-Walker equations. Because of the symmetries involved in these equations, a highly efficient computational procedure may be used. (This procedure was especially useful in the early days of computers when computational power was extremely weak and yet very expensive.) The prediction error operator is the required deconvolution operator.

(3) The final step (namely the deconvolution itself) is to convolve the deconvolution operator with the seismic trace. Note that the

"deconvolution" of the trace is accomplished by "convolving" the trace with the "inverse operator," i.e. with the deconvolution (or prediction error) operator. The result of the deconvolution is the prediction error series. The prediction error series approximates the required reflectivity function within the given time-gate.

All of the above, of course, holds within the limitations of statistical errors imposed by noise, computational approximation, and the finiteness of the data, and within the limitations of specification errors imposed by the model. The success of the method of deconvolution depends largely upon the validity of the basic hypotheses as to the minimum-delay nature of the section multiple-train waveform and to the random uncorrelated nature of the reflectivity function within the specified time gate. The power of the method of deconvolution rests in the fact that it is a stable and robust method in which the data required is the received seismic trace. The method of deconvolution is used successfully on a day-to-day basis to deconvolve field records in all seismic environments, both land and marine. The general success of the method shows that the basic hypotheses are valid over this wide range of field situations and operating conditions.

Exercises

1. Empirical studies indicate that the reflectivity is not entirely a white-noise process. Spectral properties of reflectivity functions derived from a worldwide selection of sonic logs indicate that the reflectivity is closer to blue-noise. The autocorrelation of blue-noise has a significantly large negative lag value following the zero lag. This is not the case for autocorrelation of white noise. The positive zero-lag peak followed by the smaller negative peak in the autocorrelation of the impulse response arises from the blueness of the spectrum. The positive peak and the adjacent, smaller negative peak of the autocorrelation characterize a fractional derivative operator, which produces a ramp effect in the power spectrum.

Chapter 3. Seismic migration

Gauss: Arc, amplitude, and curvature sustain a similar relation to each other as time, motion, and velocity, or as volume, mass, and density.

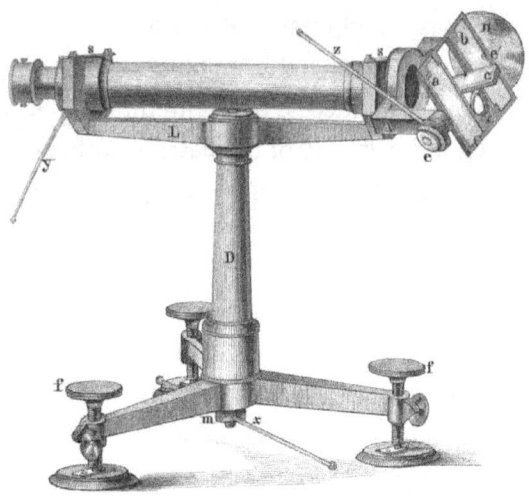

Fig. 1. Gauss heliotrope

In 1818 Carl Friedrich Gauss conducted a geodesic survey of the state of Hanover. His survey would tie into the existing Danish grid. Gauss took measurements during the day and reduced them at night, making use of his extraordinary mental skills in the calculations. Because of this work, Gauss invented the heliotrope. See Fig. 1. This instrument worked by reflecting the Sun's rays using a design of mirrors and a small telescope. In addition, Gauss developed the mathematics for statistical error analysis, and he utilized probability analysis and hypothesis testing. The normal probability curve came to be known as the Gaussian curve.

In 1820 Gauss devised a method to signal extraterrestrial beings by constructing an immense right triangle with its three squares on the surface of the Earth. See Fig. 2. This representation of the Pythagorean Theorem would be constructed on the Siberian tundra, and made up of vast strips of pine forest forming the borders of the triangle. The interior of the triangle and exterior squares would be fields of wheat. In this way extraterrestrial aliens would know about the existence of intelligent life on Earth. Such was the power of the Pythagorean Theorem. In his interest in finding a method to contact extraterrestrial life, Gauss did invent a way to employ amplified light using the heliotrope to signal supposed inhabitants of the Moon.

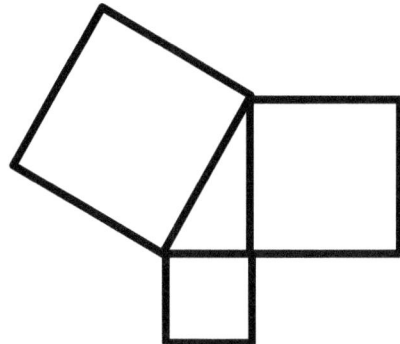

Fig. 2. Pythagorean Theorem

Hologram and wavefront reconstruction

There are three main processing operations in seismic data analysis, namely deconvolution, stacking, and seismic migration. In the preceding chapter we saw that deconvolution is an operation concerned with the time axis. Both stacking and migration are operations concerned with the spatial axes.

The objective of "3D seismic prospecting" is the creation of accurate three-dimensional computer images of the subsurface. These images show features such as the different faults and rock types as well as the distribution of porosity. The 3D images can be sliced in any direction to produce two-dimensional views, including cross-sections (vertical views or profiles) and maps (horizontal views).

The seismic industry has developed the powerful method of spatial sampling known as "multiple coverage." Let us envisage an irregular 2D array of sources placed on the surface. Let us also envisage an irregular 2D array of receivers placed on the surface. In seismic data acquisition, different combinations of source and receiver points are used to record the various seismograms. All of these seismograms are subjected to digital signal processing to produce a 3D image of the subsurface.

The 3D image fills a "3-dimensional volume" of data on an evenly-spaced 3D grid. A "3-dimensional volume" is referred to as a cube. The cube can be sliced in a vertical dimension to create a 2D profile. It can be sliced on a horizontal plane to create a depth map. The volume can also be cut along curved interfaces so as to depict the sedimentary layering.

The term "4D seismic prospecting" refers to a sequence of 3D sets with identical spatial configurations that are shot at different times for the purpose of examining the change in a reservoir over time.

The term "4-component seismic detection" is something else. It means the type of geophones used detect not only earth vibrations in the vertical direction, but also earth vibrations in the two horizontal directions, as well as pressure vibrations..

A major development in one science often produces significant impacts on other sciences. Such is the case of wavefront reconstruction or holography developed in the science of optics. Before we discuss its influence on geophysics, let us review the major points about holography as used in optics.

As we know, an ordinary photograph is a permanent record of the appearance of a field of view. When we take a photograph, we focus the camera lens on a particular plane in the field of view, and only the appearance on this plane is recorded. To a greater or less extent, all other planes are out of focus. The reason is that a photograph is a record of the intensity distribution of light. If we think of a light wave that is a sinusoidal wave of a given frequency, amplitude, and phase, then its intensity is proportional to the square of its amplitude. Thus a photograph lacks all phase information. However, if one could record both the amplitude and phase distribution on any given plane, then it is possible (according to the Kirchhoff integral theorem) to obtain the amplitude and phase distribution on any other plane. In this way, we could make a complete investigation of the field of view on any plane. In simple terms, holography represents a method of recording both amplitude and phase on a plane (the hologram), and thus from the hologram the entire picture in three dimensional space can be obtained (wavefront reconstruction).

The hologram was invented and named by Dennis Gabor in 1948 [7]. A hologram is a two-dimensional photographic plate that allows us to see a faithful reproduction of a scene in three dimensions. A regular photograph records only intensity (i.e. square of amplitude) whereas a hologram preserves both amplitude and phase. It is the phase that makes possible the scene in three dimensions. The word hold in Greek means whole, and the word gram in Greek means message. A hologram contains the whole message, or entire picture.

Let us now discuss the principle of the hologram. The principle is the same as that used in transmitting messages over the radio, except that time functions are replaced by spatial functions. A radio station broadcasts a carrier wave (a sinusoidal wave of given frequency, amplitude, and phase), and information such as speech and music is transmitted by amplitude modulation (AM) or frequency modulation (FM) of the carrier wave. The radio carrier wave is a time function. The carrier wave that appears on the hologram is a spatial sinusoidal wave. The presence of an object results in the modulation of the carrier. A hologram is recorded photographically so only an intensity distribution is recorded. However, because the information is in the form of a modulated carrier wave, we will see that the hologram records unambiguously both the amplitude and phase of the light due to the object. Thus the hologram contains the required information to reconstruct a three dimensional picture of the object.

In an ordinary camera a lens is used to form an image of the object on the plane of the photograph film. See Figure 10. Light reflected from a given point on the object is directed by the lens to the corresponding point on the film. Thus there is a one-to-one relationship between points on the object plane and points on the photograph film. Moreover, all the light that reaches the film comes from the object. There is no secondary source, i.e. there is no carrier wave. Let us now compare this situation with holography. In making a hologram no image-forming lens is used. Thus each point on the object reflects light to every point on the hologram plate. Thus there is a one-to-many relationship between points on the object and the points on the hologram plate. Thus every part of the hologram plate is exposed to light reflected from every part of the object. In addition all the light that reaches the hologram plate comes from two places, namely from the object and from part of the beam that is used to illuminate the object. This illuminating beam produces the carrier wave on the hologram. The illuminating beam is coherent light as produced by a laser.

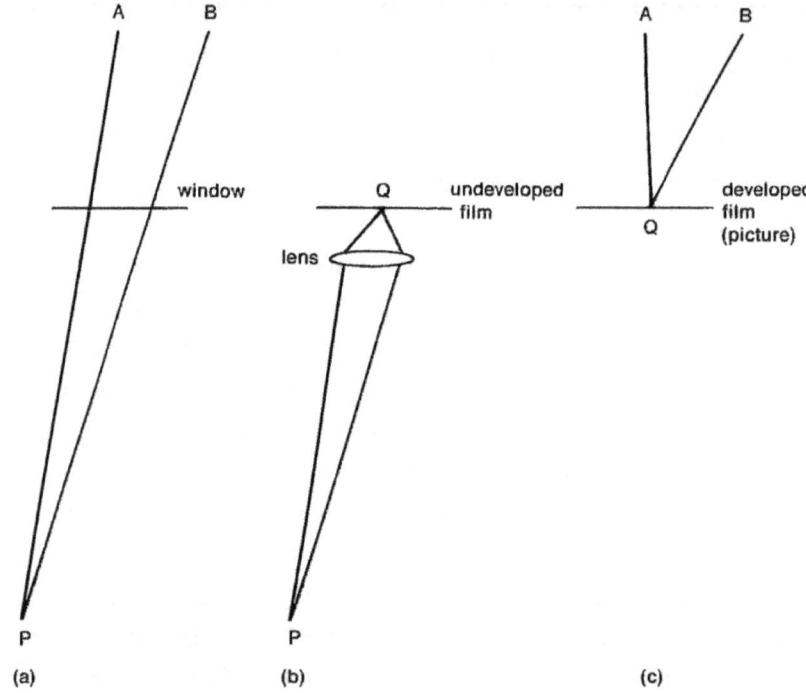

Fig. 3. A picture is not a window

See Fig. 3.

(a) Looking through a window, we see point P in different perspectives from A and B.

(b) In taking a photograph all the light rays from P are directed by the lens to point Q on the photograph.

(c) Looking at the picture, we see point Q in the same perspective from both A and B.

Chapter 3. Seismic migration

A hologram is a recording of the interference pattern resulting from the combination of two sets of wavefronts. One set of wavefronts is from the reference beam (the carrier wave), and the other set of wavefronts is from the light reflected from the object. The developed hologram thus represents the carrier wave modulated by waves from the object. If we look at a hologram we see no recognizable image. The hologram is dark where the object wave and the reference wave arrive in phase, and light where the object wave and reference wave arrive out of phase. Thus the intensity of the hologram corresponds to a given phase difference between the object waves and the reference waves, and is unaffected by a change in sign of that difference. A hologram is a photograph of microscopic interference fringes, and appears as a hodgepodge of whirly lines.

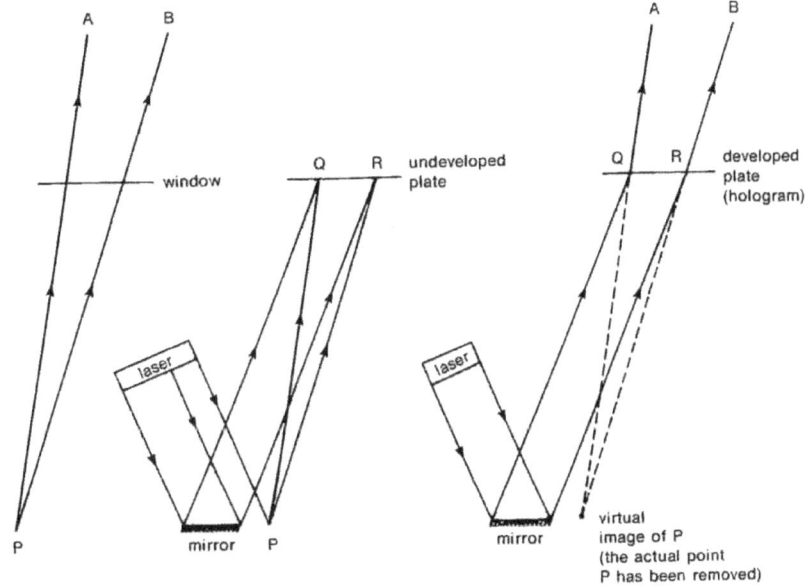

Figure 4. A hologram is a window

See Fig. 4. When the hologram is placed in a beam of laser light (with the object gone) the light rays are bent by the hologram to produce rays identical to the original rays reflected by the object. When view by the eye the bent (or diffracted) rays produce the same effect as the original reflected rays. When we look through the hologram we see a full realistic three-dimensional virtual image as if we were viewing the object through a window. When we move our eyes and look down the sides of the object, and when we lower our eyes and look underneath the object, parallax is evident as in real life. The entire wavefield on our side of the hologram has been reconstructed by the illumination of the hologram by the laser light. We see the object as a virtual image, even though the original object is no longer present.

Seismic section and wavefront reconstruction

(a) Looking through the window, we see point P in different perspectives from A and B.

(b) In making a hologram, the light rays from P are directed at all points on the plate, such as points Q and R.

(c) The developed hologram diffracts (i.e. bends) the laser rays so as to form a virtual image of P. (The actual point P has been removed beforehand.) Thus looking at the hologram we see the virtual image in different perspectives from A and B.

Seismic surveys are taken at points on a grid on the surface of the earth in order to determine the underlying three-dimensional structure. However, for the purposes of exposition it is easier to consider the case where seismic data is taken along a single surface line, and regard the underlying earth structure as two dimensional. In the corresponding two dimensional case, a hologram plate would be a line, and the hologram values would be the intensity of the modulated carrier wave along this line. We can think of the intensity recorded at a certain point on the hologram in the following way. During exposure of the photographic plate the point in question experiences vibrations in the form of a digital signal representing the received wave motion. The photographic film does not record the digital signal, but records the intensity of the digital signal. In terms of digital signal analysis, the intensity is the zero-lag autocorrelation coefficient. Thus the hologram is a record of the zero-lag autocorrelation coefficients of the digital signal received at each point on the hologram. Because the intensity is the time-average of the squared amplitudes of the received digital signal, it follows that the intensity is intrinsically positive. Because vast amounts of information are recorded on a hologram, the film used for a hologram must have a resolving power much greater than ordinary fine-grain photographic film. The information contained on our one-dimensional hologram (i.e. a hologram where the plate is simply a line) is enough to reconstruct the wavefield in a plane. The hologram is indeed a remarkable invention.

Let us now turn to the seismic case. First of all in the seismic case we are using acoustic (sound) waves instead of electromagnetic (light) waves. Whereas the laser waves are extremely narrow-band waves, and are in fact, almost pure sine waves, the acoustic waves are broad-band. That is, acoustic waves look like the digital signal familiar to every time-series analyst. In seismic prospecting a source of acoustic waves is directed into the earth. However, we cannot place a high quality reflecting mirror within the earth in order to generate a reference beam as we do in holography. Only the waves reflected from geologic objects in the subsurface (i.e. the geologic structure) are returned to the recording line on the surface. This seismic recording line corresponds to the hologram. We are now ready to state the essential difference between holography and seismology. On each point on the hologram the photographic film records only the zero-lag autocorrelation coefficient of the received digital signal. On each point of the seismic line, the seismic instruments record the entire received digital signal. Thus the received seismic data is two dimensions, one dimension corresponding to the horizontal surface line and the other dimension corresponding to time. From these data, we want to reconstruct the two-dimensional subsurface structure, one dimension corresponding to the horizontal surface line and the other dimension corresponding to depth. The seismic recording instruments are able to record the entire surface wave motion as a function of time, and not just the intensity (the time-average of the squares of the amplitudes). Because the depth that a wave penetrates and its travel time are related by velocity of the seismic waves, there is a correspondence between the depth axis and the time axis. Thus the recorded seismic data (called the seismic section) as a function of horizontal coordinate and time is a rough picture of the cross-section of the earth which is a function of horizontal coordinate and depth. In this sense the seismic section is like a picture, albeit distorted, of the subsurface geologic structure. The great concentration of information as in a hologram is not present in a seismic section.

Reflection seismology was developed in the 1920's. By the 1930's it was being used on a large scale for oil exploration. Various methods of transforming the seismic section in order to yield a better picture of the subsurface structure came into prominence in the 1940's. These general methods were labeled with the name *seismic migration*, as the methods would move (or *migrate*) the apparent position of subsurface interfaces appearing on the seismic time section to their true position in depth. A unified theory of seismic migration methods based upon the wave equation in the form of wavefront curves and diffraction patterns was given by Hagedoorn. His work made much use of the theory of ray paths. As computers became more and more powerful, wave equation methods became predominant in seismic wavefront reconstruction and imaging.

The first requirement in the presentation of any data processing method is a description of the model. Often in the historical development of the method the model does not evolve first, but only later as the practical effectiveness of the method becomes better understood. The model serves as an aid to the understanding of the method. The basic model for conventional seismic migration techniques is called the distributed source model. Although this model is quite intuitive and so basic that it is self-evident, it does help clarify the concept of migration. Everything we see except the sun, or an electric light bulb, or the like, is by reflection. We see the moon because the sun's rays are reflected from the moon. However, in the data processing and imaging that takes place in our mind, we think of the moon as the source of the light. In the same way the geologic objects that we hear from underground (for seismic waves are sound waves in the earth) in a reflection seismic survey come to us by reflection of the source energy that we initiate. However, in the seismic data-processing scheme of migration, it is more convenient to think of the underground objects as being the source of their own energy. Because these underground objects are in the form of sedimentary layers, which are distributed according to the geologic structural characteristics of the region, we call this model the distributed source model.

Let us now describe the source of the energy. Reflection seismology is an echo-ranging technique. Other such techniques are radar, sonar, and various ultrasonic methods. The convention source for echo ranging is a very short and very intense signal, which is called an impulse or spike. The signal travels to the inaccessible object and then returns by reflection. The two-way travel time, i.e. the time to the object and the time back to the receiver position, is recorded. If the source and receiver are at the same point (such as in radar where the transmitter and receiver use the same dish), then the one-way travel time is one-half the two way travel time. In seismic prospecting, the sources and receivers are at various points, but the data is processed to give what is called a common midpoint (CMP) section. This section can be regarded as one for which the source and receiver for each digital signal (seismic trace) is at the same point. If we divide the time scale on the CMP section by two, we obtain one-way time. We can now regard the CMP section as being generated by the distributed source model. This model, which is also called the exploding reflector model [12], can be described as follows. Each subsurface reflector (i.e. each interface between two geologic strata) is considered to be composed of a continuous distribution of sources, with the source magnitude given by the reflection coefficient of the interface. All the sources are set off at the same instant (which is time zero) and the waves travel to the surface where they are recorded. Each surface point thus records a digital signal (called a seismic trace). All the traces make up the CMP seismic section.

Suppose now that the earth consists of a series of flat horizontal beds, much like a layer cake. Sources are distributed on each interface, and at time zero these sources are set off as impulsive signals. These spikes travel straight up and identical digital signal (i.e. the same seismic trace) will be recorded at each surface point. Let us describe this common digital signal. It consists of a series of spikes. The first spike to arrive comes from the first (i.e. shallowest) interface; the second spike comes from the next deepest interface, and so one. When we put all these identical digital signal side by side to make up the seismic section, all the events due to the first interface line up in a straight line. Likewise all the events due to the second interface line up in a straight line. Thus the seismic section as recorded on the surface of the earth is a true image of the cross-section of the earth. So much for the case of flat horizontal beds.

Let us now look at a dipping model. Suppose that the earth consists of a series of flat beds, all parallel, but not horizontal but dipping. At time zero the impulsive sources on the interfaces are set off. By Fermat's principle of least time the ray paths are normal to the interfaces, and hence the waves do not travel straight up but travel in sloping paths along the rays. The recorded seismic traces again consist of spikes corresponding to the interfaces. When we put all these digital signal side by side to make up the seismic record section, all the spikes do line up in straight lines corresponding to the interfaces. However the slopes of the event lines on the seismic section are not the same as the slopes of the underground interfaces. The seismic section is still an image of the subsurface structure, but a distorted image. The purpose of seismic migration to go back from this distorted image to a true image of the subsurface layers. So much for flat dipping beds.

Oil and natural gas are usually found in sedimentary basins. The sediments (during geologic time) were laid down as deposits in shallow seas, and hence the sedimentary rock layers in the earth do tend to be flat and horizontal. However, in many of the interesting areas where petroleum is formed, the various strata have been folded, tilted, eroded, and faulted. As the geologic structure deviates more and more from that of flat horizontal layers, so does the seismic section recorded at the surface become more and more distorted as an image of the geologic structure. It is for this reason that the data processing technique known as migration is so important in exploration. Migration removes the distortion by projecting the wave motion backward to the sources distributed on the geologic interfaces. In order to achieve this background projection, we must make use of the wave equation.

Wave equation as the basis for seismic migration

Wave motion represents a phenomenon which unifies almost all of the physical sciences. Wave motion is described by various types of wave equations. The physical experiment involves the propagation of waves from subsurface sources to the receivers on the surface of the ground. Time goes in the positive direction as the waves travel upward. Given the sources the wave equation can be evoked to compute the received wavefield (i.e. the seismic section) measured on the surface. This physical process can be simulated within a computer. The computer memory can be used as a representation of physical space, and time evolves from initial time zero. Thus we would start with the initial values representing the sources distributed on the interfaces, and let the computer simulate the wavefield as time increases. This type of simulation corresponds to what occurs in nature. The wavefield evolves according to the wave equation in a unique way depending upon the initial conditions. Generally small errors in the initial conditions as well as in the model specification do not propagate in an ever-increasing way. Thus given the initial source distribution which represents the structure of the earth, we can compute the wavefield everywhere. In particular we can compute the wavefield as it appears on the surface of the earth. This surface wavefield is the received seismic section. The process of obtaining this wavefield from the subsurface structure represents wavefield construction.

In geophysical exploration, on the other hand, we are faced with what is called an inverse problem. We have the digital signal data (the seismic traces) that we have measured at the surface of the ground, and we want to determine the underground structure. We are given the surface data and we want to extrapolate the wavefield downward in depth and backward in time. That is, in the inverse problem we run the clock backward and let the waves return from where they came. When the backward running clock reaches time zero, we have reached the distributed sources which delineate the subsurface structure. Let us summarize. We take the surface data and from it reconstruct the wavefield for prior times and for various depths. The values of this reconstructed wavefield at time zero gives the distributed sources.

These distributed sources lie on the reflecting interfaces, and hence depict the geologic structure. Generally speaking, inverse problems do not have the nice stability and uniqueness properties as do direct problems. However when we use the wave equation to extrapolate backward in time and down into the earth, stability and uniqueness can be attained suitable mathematical and numerical techniques.

Seismic data processing makes use of efficient, robust, and stable methods of solving the geophysical inverse scattering problem. The solution of this problem represents the transformation of surface data into subsurface data. The surface data is obtained by geophysical prospecting methods. Computers then perform the operations of deconvolution, stacking, and migration and the final output in a depth section of the earth which is used to determine favorable drilling sites for oil and gas. In this paper we have treated deconvolution, stacking, and migration as separate operations; ideally, they would be incorporated with each other in various interrelationships. In a paper of this length we have treated only some of the main aspects of seismic data processing; there are many other aspects including static and dynamic corrections, adaptive and self-correcting modelling, and velocity determinations which are of vital importance in oil and gas exploration.

Chapter 4. Wave motion

Gauss: I have had my results for a long time: but I do not yet know how I am to arrive at them.

Introduction

The seismic waves which travel through the rock layers in the earth are mechanical waves, as these waves involve the actual motion of the rock particles. Mechanical waves are of two fundamental types: longitudinal and transverse. In a longitudinal wave, the oscillating particles of the medium are displaced parallel to the direction of propagation (i.e., the direction of energy transmission) of the wave. In a transverse wave, the particles are displaced in a direction perpendicular to the propagation direction. When a steel rod (i.e., a slim straight bar of metal) is struck on one end by a hammer, a wave pulse in the form of a longitudinal compression of the rod travels down its length. Because these waves travel through the entire body of the rod, they are body waves. If the rod is struck periodically, a succession of such pulses, known as a wave train, travels down the rod. The concept that sound propagates through the air as longitudinal waves was familiar to Aristotle. The water waves that come from the point where a stone is dropped into a quiet pond are transverse waves. Since these waves are confined to the water layer close to the surface, they are surface waves. Seismic waves that travel through the rock layers are body waves, and they can be either longitudinal waves or compressional waves. In addition, there are various types of seismic surface waves that can be identified on seismograms.

The phenomena perceived by our eyes as light and with our ears as sound are propagated as wave motion. Their motion occurs not in the two-dimensional surface of a pond, but in three-dimensional space. However, many of the properties of all wave motion can be seen by studying the familiar water waves. A stone produces water waves. The waves move out in circular rings at a constant speed. This wave speed is called the velocity of propagation, and is denoted by v. The waves themselves have crests and troughs, that is, points where the water level is elevated and points where the level is depressed. The water surface undulates rhythmically between crest and trough, crest and trough. The distance between successive crests, or alternatively the distance between successive troughs is called the wavelength. It is usually denoted by the Greek letter λ (lambda). As the waves travel past a fixed point on the surface of water, they cause a vertical up-and-down motion of the water at this given point. This up-and-down motion repeats itself in time in a periodic manner. The number of times per second that this up-and-down motion repeats itself is called the frequency of the wave. It is denoted by f, or by the Greek letter v(pronounced nu). There are three fundamental aspects of wave motion: (1) the velocity v with which the wave travels onward or propagates, (2) the distance between crests (or between troughs), that is, the wavelength λ, and (3) the frequency f with which the medium pulsates to and fro.

Taylor and Fourier

The two best known types of infinite series in mathematics are the Taylor series and the Fourier series. It is inconceivable to have any serious development of mathematics without them; they are of fundamental importance in applied mathematics. A careful study of these two types of series is certainly warranted for every engineer and scientist, and especially so for geophysicists. We can approach the subject historically, principally by examining the lives of the people whose discoveries formed the basis for our present mathematical knowledge. There are many different stories, intertwined and related in different ways, with the elements of chance and design intermixed. But, there is one story, largely untold, which certainly qualifies as one of the most interesting in the history of science. We tell that story, the narrative of Taylor and Fourier which is held together by the magic of the vibrating string. The story starts late in the seventeenth century with the invention of calculus. All of the scientific developments in physics were made possible by this epic invention.

Chapter 4. Wave motion

From ancient times, it was understood that material objects are the sites of continual motion. Things move from one place to another. They undergo transformations. They change to other things and move to other places. Nothing remains constant forever. There are many kinds of change. Changing position is called motion. However, motion is always relative to something. A tree rooted to Earth is not moving relative to earth. But, the Earth is in orbit about the sun, and in turn the sun is in orbit in the Milky Way galaxy. Thus, relative to an observer in space, the tree is moving. When describing motion, some system must first be specified. The law of inertia, which was discovered by Galileo and included as Newton's first law, states that every material object continues in its state of rest, or in uniform motion in a straight line, unless it is compelled to change that state by forces impressed on it.

Changing position can take place in many ways. The motion may be rectilinear (that is, on a straight line), it may be curvilinear, or it may be oscillatory. Further, motion may occur with constant velocity, or with changing velocity, in which case it is said to be accelerated motion. Motion is most often and easily conceptualized by setting up a coordinate system, for example, a Cartesian representation. This technique, very familiar in the construction of graphs, was invented by the French philosopher and mathematician Rene Descartes (1596–1650). The Cartesian axes constitute the frame of reference which gives meaning to statements about motion of objects in a system. In describing the motion of an object, we are concerned with two related changes: (1) in position and (2) in time. The distance traveled is usually described as a function of elapsed time, but another equally valid point of view would be that time is a function of distance.

Function is a term basic to mathematics and science. It is the name of any mathematical expression that describes a relation between variables, that is, between linked values of things that can change. In accordance with a definition that the great mathematician Leonhard Euler (1707–1783) gave in 1749, a function is often explained as a variable quantity that is dependent on another variable quantity. The mathematical conception regarding the nature of functions, expressions that fix the relationship between variable quantities, has a long evolution. Serious investigation of functions did not begin until the seventeenth century when two great advances were made in quick succession. One was analytic geometry by Descartes and the other was calculus.

In effect, calculus is a mechanical procedure with the required notation to deal with infinitesimals. Of course, there was a large history of such work without calculus, but the pre-calculus period ended when Barrow introduced the differential triangle and proved the fundamental theorem of calculus. Isaac Newton and Gottfried Wilhelm von Leibniz (1646-1716) each then developed his own version and system of notation for calculus, but that of Leibniz survived and that of Newton did not. The new and extremely powerful method of calculus made more exact knowledge of functions and their properties supremely important. At that time, few even understood calculus, but educated people recognized that calculus was important. It not only opened ways to calculate velocities and other rates of change, but it also opened innovative avenues in all sciences and lead to the development of new scientific instruments.

Tayler series

The two best known types of infinite series in mathematics are the Taylor series and the Fourier series. It is inconceivable to have any serious development of mathematics without them; they are of fundamental importance in applied mathematics. A careful study of these two types of series is certainly warranted for every engineer and scientist, and especially so for geophysicists.

Newton's second law of motion, which says that force is the product of mass and acceleration, is usually expressed in the simple formula $F = m\,a$. However, this is really a second order differential equation because acceleration is the second derivative of the function which relates distance and time. If x is distance and t is time, then Newton's second law can be written as the differential equation

$$F = m\,\frac{d^2y}{dx^2}$$

Chapter 4. Wave motion

In the seventeenth century and into the eighteenth century, it became commonplace to assume that any function describing a relationship between physical variables would be differentiable. The idea that such a function could change in a capricious or random way, and thus would not be differentiable at all points, did not enter into mathematical thinking at that time. It is now known that functions are not restricted and some can act capriciously. However the mistaken idea of universal continuity and unlimited differentiability most likely prompted Brook Taylor (1685–1731) to investigate the possibility that a function could be expressed in terms of its derivatives. Taylor found this was true and in 1715 introduced analytic functions into mathematics.

The core of his discovery was the Taylor series, i.e., that an analytic function can be expressed as the summation of other functions. More precisely, the Taylor series expands the function $f(x + h)$ at the point $x + h$. It is assumed that the function is differentiable to all orders in the neighborhood of the point x. The infinite series has coefficients given by the function $f(x)$ and its successive derivatives

$$f'(x), f''(x), \cdots$$

at the given point x. The Taylor series can be expressed as

$$f(x + h) = f(x) + h f'(x) + \frac{h^2}{2} f''(x) + \cdots$$

This is an astonishing result because the definition of the derivative of any order at the point x requires nothing more than knowledge of the function in an arbitrarily small neighborhood of this point. The Taylor series therefore establishes that the shape of the function at any finite distance h from the point x is uniquely determined by the function's behavior in the infinitesimal vicinity of the point. This property of Taylor series implies that an analytic function has a strong interconnected structure which makes it possible to predict precisely what will happen at any point a finite distance from a certain point by merely studying the function's behavior in a small vicinity of point. This property is extremely valuable in mathematical analysis because, although it is not common to all functions, it is shared by many of the most useful and regularly encountered, such as polynomials, the sine and cosine functions, and the rational functions (away from their poles).

Pendulum

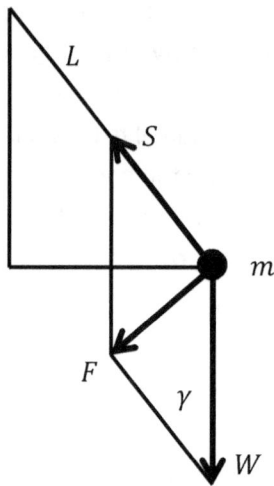

Fig. 1. Net force on a pendulum bob

Consider the motion of a simple pendulum, ideally just a point mass m at the end of a rigid and massless support of length L. See Fig. 1. Can it be determined whether the simple pendulum does indeed execute simple harmonic motion? If so, can a formula for the period T of this pendulum be derived? The first of these aims is fulfilled if it can be established that the path of the bob is (nearly enough) a straight line and that the net force on the bob is proportional to the displacement and always directed to the center (equilibrium) position. If the motion of the bob is restricted to suitably small angles, the straight line condition is approximately met (and x = arc length). The net force (F in the figure) is drawn as the resultant of the tension S in the weightless string or rod, and of the weight W of the bob. Note that F must be perpendicular to the string, and thus to S. (If this point offers difficulty, try to think what would have to happen if F were not perpendicular to S. Any component parallel to the string will extend it, move it, or break it.) Therefore, from the drawing, F has the magnitude equal to $\sin \gamma$. But $\sin \gamma = x/L$ so F has magnitude x/L, or F is proportional to x (in magnitude). The drawing shows also that F is always directed to the equilibrium position which means that F and x have opposite sign. As we know, a force which has algebraic sign opposite to that of x is called a restoring force. Thus we have $F = -W(x/L)$. Hence, the bob will execute simple harmonic motion for small angles of swing. As we know $F = -\alpha x$ and $\omega = \sqrt{\alpha/m}$ for simple harmonic motion. Thus for the present problem

$$\alpha = \frac{W}{L} \quad \text{and} \quad \omega = \sqrt{\frac{W}{mL}}$$

Therefore the period of a pendulum is

$$T = \frac{2\pi}{\omega} = 2\pi\sqrt{\frac{mL}{W}}$$

Note that $W = mg$, where g is the acceleration due to gravitation. It follows that the period of a pendulum is given by

$$T = 2\pi\sqrt{\frac{mL}{mg}} = 2\pi\sqrt{\frac{L}{g}} \qquad (1)$$

This formula holds as long as the swing is not too large, say less than six degrees of arc. This is a somewhat unexpected result, for it implies that except for air resistance, a heavy and a light pendulum bob will both swing side by side in step if suspended from equally long strings. Also, the period is exactly the same for a medium-sized swing as for a very small one. This is referred to as the *isochronism* of the pendulum, a phenomenon which Galileo is reputed to have discovered experimentally when a young student.

Incidentally, in the equation (1) the two expressions for mass are not conceptually identical: m in the numerator refers to inertial mass, m in the denominator to gravitational mass. Only if these two have the same numerical value, can we cancel them. Conversely, if equation (1) is found to hold experimentally, as it does, then we can deduce that these two values are really identical. Newton himself made pendulum bobs of a great variety of substances, and so found that equation (1) holds, and that therefore the two types of masses are equivalent.

Let g_1 and g_2 refer to the values of the gravitati0onal constant g in Paris and Cayenne respectively, and let T_1 and T_2 refer to the corresponding periods respectively for the same clock. Thus,

$$4\pi^2 L = g_1 T_1^2 = g_2 T_2^2$$

When he wrote the *Principia*, Newton knew that in 1672 Jean Richer had taken a pendulum clock from Paris to Cayenne, French Guiana, to help in astronomical observations. There the clock lost 2 1/2 minutes each day, so he could calculate T_2 from T_1. From g_1 for Paris, Newton thus found g_2 for Cayenne. It was this observation which drew general attention to the distinction between the concepts of mass m and weight, which would be mg_1 in Paris but would be mg_2 in Cayenne. Until then mass and weight were not distinguished as separate entities.

Wave motion

In classical times, there were four elements Earth, Water, Air, and Fire frequently; and sometimes a fifth element or quintessence (after "quint" meaning "fifth") called ether. A wave is a traveling disturbance. Waves of one form or another can be found everywhere. Seismic waves travel through the earth; ocean waves travel through the water; sound waves travel through the air; heat waves travel through fire, and electromagnetic waves travel through the ether. The ether was originally regarded as a material substance but after the Michelson-Morel experiment the ether is regarded as vacuum (i.e., the absence of a material substance.

Much of the historical understanding of wave motion has come from the study of the sound waves that make up music. The Greek mathematician Pythagoras (570 - 490 BC) founded his school in the Greek colony of Kroton in southern Italy. Pythagoras was the first to advance the notion of heliocentricity (planets orbiting the sun), and was also earliest advocate of the Solar Calendar, which he thought to be better than the lunar calendar in use throughout the Mediterranean region. Although Pythagoras demanded silence about his esoteric work, his thinking played a noteworthy role in the creation of western civilization. Pythagoras used numbers to model everything in the physical world, including music. Pythagoras hypothesized that there was

Chapter 4. Wave motion

a connection between waves and sound, and that vibrations, or disturbances, must be responsible for sounds. By analysis of the sounds produced by vibrating strings, Pythagoras determined mathematical relationships between the lengths of strings that make harmonious tones. Pythagoras observed that vibrating strings produced sound, and worked to determine the mathematical relationships between the lengths of strings that made harmonious tones. His discovery of the harmonic progression of simple whole numbers represented work influential in the creation of modern science.

The study of acoustics has contributed much of the understanding of wave motion. Pythagoras left no written record of his work; it was against his esoteric principles. Pythagoras realized that there is a connection between waves and sound. In observing how a vibrating string produces sound, he discovered mathematical relationships between the lengths of strings that made harmonious tones. Our tuning system and scale (with the division of the octave into 12 steps) were constructed to incorporate the intervals originally discovered by Pythagoras.

Pythagoras investigated the harmonic relationships between vibrating strings of different lengths using the monochord. The monochord, said to have been invented by Pythagoras and illustrated above, was a rectangular box-shaped instrument, with two fixed bridges (shown in black) across which a single string passed. One end of the string was wound around a peg while the other end was looped over a small metal pin driven into the frame of the box. The string's tautness was adjusted by turning the peg. The string could be bowed or plucked and a moveable bridge (shown in red), lying between the two fixed bridges, divided the string into two shorter lengths. When the moveable bridge was placed exactly half way between the fixed bridges, the two shorter lengths would sound the same note; the ratio of the two 'half-lengths' was then 1:1. If the moveable bridge was placed one third of the distance between the fixed bridges, so that the string was divided in the ratio 2:1, the two strings would now sounded one octave apart, the longer string sounding the lower note. In this way it was found that the ratio 2:3 (called 'hemiolic', which means the whole exceeded by its half) was associated with the interval of a fifth and 3:4 (called 'epitrite', which means the whole exceeded by its third) gave an interval of a fourth.

The discovery that musical harmonies are themselves numerical ratios, that the 'consonant' harmonies are represented by ratios of the four integers 1, 2, 3 and 4 (for which Pythagoras had a particular regard), and that a string shortened to half of its original length produces a note which is one octave higher was seen as confirming their metaphysical idea that number is the actual essence of things.

The notion of a wave originates with water waves. A wave is an oscillatory disturbance that travels away from a source. A wave transports no discernible amount of matter over large distances of propagation. The awareness that sound is a wave developed from observations of water waves. The explanation was in line with Aristotle's (384-322 B.C.) account that sound is generated by a source, "thrusting forward in like manner the adjoining air, such that the sound travels unaltered in quality as far as the disturbance of the air manages to reach."

The analogy with water waves led to the realization that (1) air motion associated with musical sounds is oscillatory, (2) sound travels with a finite speed, and (3) sound bends around corners (diffraction). In the seventeenth century, some argued that sound is due to a stream of so-called atoms" emitted by the sounding body. This viewpoint did not last. However the encounter between ray theory and wave theory was prevalent optics. When ray concepts are used to explain acoustic effects, they are regarded as mathematical approximations to wave theory. Geometrical optics and geometrical seismology are extremely useful in explaining wave phenomena in terms of rays.

The physics of sound has retained ideas of the ancient Greeks including the proposition that sound is a 'mechanical wave' analogous in some way to a wave traveling over water. This speculation that sound is a wave phenomenon grew out of observations of water waves. The rudimentary notion of a wave is an oscillatory disturbance that moves away from some source and transports no discernible amount of matter over large distances of propagation. The analogy with water waves was strengthened by the belief that air motion associated with musical sounds is oscillatory and by the observation that sound travels with a finite speed. Another matter of common knowledge was that sound bends around corners, which suggested diffraction, a phenomenon often observed in water waves.

Much of the current understanding of wave motion has come from the study of acoustics.

Vibrating string

At the time of Newton, it was generally assumed, because contemporary observations of natural events seemed to indicate a continuous relationship between physical variables, that all functions were continuous. This view was reinforced by the frequent use of the new tool calculus to formulate natural laws in terms of differential equations. For example, Newton's second law of motion, which says that force is the product of mass and acceleration, is usually expressed in the simple formula $F = ma$. However, this is really a second order differential equation because acceleration is the second derivative of the function which relates distance and time. If x is distance and t is time, then the familiar $F = ma$ can be written as the differential equation

$$F = m \frac{d^2 x}{dt^2}$$

As a result, it became commonplace to assume that any function describing a relationship between physical variables would be differentiable. The idea that such a function could change in a capricious or random way, and thus would not be differentiable at all points, did not enter into mathematical thinking at that time. It is now known that functions are not restricted and some can act capriciously.

The mathematician Brook Taylor (1685–1731) discussed the motion of projectiles, the center of oscillation, and the forms taken by liquids when raised by capillarity. The mistaken idea of universal continuity and unlimited differentiability most likely prompted Taylor to investigate the possibility that a function could be expressed in terms of its derivatives.

Taylor found this was true and in 1715 introduced analytic functions into mathematics.

The core of his discovery, given in his work *Methodus Incrementorum*, is the Taylor series, i.e., that an analytic function can be expressed as the summation of other functions. More precisely, the Taylor series expands the function $f(x + h)$ at the point $x + h$. It is assumed that the function is differentiable to all orders in the neighborhood of the point x. The infinite series has coefficients given by the function $f(x)$ and its successive derivatives $f'(x), f''(x),\ldots$ at the given point x. The Taylor series can be expressed as

$$f(x + h) = f(x) + hf'(x) + \frac{h^2}{2}f''(x) + \cdots$$

This is an astonishing result because the definition of the derivative of any order at the point x requires nothing more than knowledge of the function in an arbitrarily small neighborhood of this point. The Taylor series therefore establishes that the shape of the function at any finite distance h from the point x is uniquely determined by the function's behavior in the infinitesimal vicinity of the point. This property of Taylor series implies that an analytic function has a strong interconnected structure which makes it possible to predict precisely what will happen at any point a finite distance from a certain point by merely studying the function's behavior in a small vicinity of point. This property is extremely valuable in mathematical analysis because, although it is not common to all functions, it is shared by many of the most useful and regularly encountered, such as polynomials, the sine and cosine functions, and the rational functions (away from their poles).

In addition, Taylor's book *Methodus Incrementorum* also includes several theorems on interpolation. Taylor was the earliest writer to deal with theorems on the change of the independent variable. Taylor was perhaps the first to realize the possibility of an operational calculus, and he is usually recognized as the creator of the theory of finite differences. His *Methodus Incrementorum* also contains the earliest determination of the differential equation of the path of a ray of light when traversing a heterogeneous medium; and, assuming that the density of the air depends only on its distance from the earth's surface,

Taylor obtained, by means of integration, the approximate form of the curve. The form of the catenary and the determination of the centers of oscillation and percussion are also discussed. The applications of the calculus to various questions given in the *Methodus Incrementorum* have hardly received the attention they deserve. If they did, Taylor would be established as one of the great mathematicians of all time. As it now stands, Taylor is in effect hidden in the shadow of Newton.

Among the problems that Taylor discusses in the second part of *Methodus Incrementorum* are two that deal with the vibrating string. They had already been discussed in Taylor's paper "De motu nervi tensi" ("On the motion of a tense sinew"), in *Philosophical Transactions* in 1714.

The first problem is to determine the motion of a tense string; the second is given the length and weight of the string, as well as the stretching weight, to find the period of vibration. There is no evidence that Taylor had any notion of partial derivatives. However, in order to appreciate what Taylor accomplished, we will make use of partial derivatives in the discussion that we now give. Taylor concludes that at any point of the arc, the normal acceleration

$$\frac{\partial^2 u}{\partial t^2}$$

is proportional to the curvature

$$\frac{\partial^2 u}{\partial x^2}$$

This means that, for small vibrations, Taylor has in principle discovered the wave equation, now written as

$$\rho \frac{\partial^2 u}{\partial t^2} = F \frac{\partial^2 u}{\partial x^2}$$

The quantity ρ is the mass per unit length of the string and F is the tension as determined by the weight which stretches the string. He did find, however, that the motion of an arbitrary point is like that of a simple pendulum. More specifically, Taylor showed that the period is equal to

$$T = 2\sqrt{\frac{mL}{F}}$$

where L is the length of the string and m its mass. Because $m = \rho L$ we have

$$T = 2\sqrt{\frac{\rho L^2}{F}} = 2L\sqrt{\frac{\rho}{F}}$$

Let us now define the constant v as

$$v = \sqrt{\frac{F}{\rho}}$$

Then the period becomes

$$T = \frac{2L}{v}$$

This equation shows that the period T is proportional to the length L of the string. Taylor took the form of the curve to be sinusoidal.

In conclusion, Taylor made two great discoveries in reference to the vibrating string, namely that the motion satisfies equation (6) which we know today as the wave equation, and that sinusoidal curves with period given by (10) are solutions of the wave equation.

The quantity F/ρ has dimension of

$$\frac{\text{force}}{\text{mass/length}}$$

But, force is equal to mass times acceleration, so the above becomes

$$\frac{\text{mass} \cdot \text{acceleration}}{\text{mass/length}} = \frac{\text{mass} \cdot (\text{length/time}^2)}{\text{mass/length}} = \frac{\text{length}^2}{\text{time}^2}$$

Thus F/ρ has the dimension of the square of velocity. For that reason, we may call

$$v = \sqrt{\frac{F}{\rho}}$$

the velocity. Then the equation

$$\rho \frac{\partial^2 u}{\partial t^2} = F \frac{\partial^2 u}{\partial x^2}$$

becomes

$$\frac{\partial^2 u(x,t)}{\partial x^2} = \frac{1}{v^2} \frac{\partial^2 u(x,t)}{\partial t^2}$$

This equation is called the *wave equation*.

Jean Le Rond d'Alembert

In 1744, Jean Le Rond d'Alembert published his *Traité de l'Équilibre et du Mouvement des Fluides*, in which he applies his principle to fluids; this led to partial differential equations which he was then unable to solve. In 1745, he developed that part of the subject which dealt with the motion of air in his *Théorie Générale des Vents*, and this again led him to partial differential equations. A second edition, in 1746, was dedicated to Frederick the Great of Prussia, and procured an invitation to Berlin and the offer of a pension; he declined the former, but subsequently, after some pressing, pocketed his pride and the latter.

In 1747, he applied the differential calculus to the problem of a vibrating string, and arrived at the *wave equation* (with $v = 1$):

$$\frac{\partial^2 u(x,t)}{\partial x^2} = \frac{\partial^2 u(x,t)}{\partial t^2}$$

This is the wave equation as we know it today. d'Alembert succeeded in showing that it was satisfied by

$$u(x,t) = f(x+t) + g(x-t)$$

where f and g are arbitrary functions. This solution was published in the *Transactions of the Berlin Academy* for 1747. The proof begins by saying that, if we define p and q as

$$p = \frac{\partial u}{\partial x} \text{ and } q = \frac{\partial u}{\partial t}$$

then we have the exact differential

$$du = p \, dx + q \, dt$$

In these quantities the wave equation becomes

$$\frac{\partial q}{\partial t} = \frac{\partial p}{\partial x}$$

Therefore

$$p\,dt + q\,dx$$

is also an exact differential; denote it by dv. Therefore

$$dv = p\,dt + q\,dx$$

Hence

$$du + dv = p\,dx + q\,dt + p\,dt + q\,dx = (p+q)(dx+dt)$$

And

$$du - dv = p\,dx + q\,dt - p\,dt - q\,dx = (p-q)(dx-dt)$$

Thus $u + v$ must be a function of $+t$, and $u - v$ must be a function of $x - t$. We may therefore put

$$u + v = 2f(x+t)$$

and

$$u - v = 2g(x-t)$$

By adding, we can eliminate the v terms, leaving

$$u = f(x+t) + g(x-t)$$

which is the d'Alembert solution of the wave equation.

The chief remaining contributions of d'Alembert to mathematics were on physical astronomy, especially on the precession of the equinoxes, and on variations in the obliquity of the ecliptic. These were collected in his *Système du Monde*, published in three volumes in 1754. During the latter part of his life, he was mainly occupied with the great French encyclopedia. For this he wrote the introduction, and numerous philosophical and mathematical articles; the best are those on geometry and on probabilities. His style is brilliant, but not polished and faithfully reflects his character, which was bold, honest, and frank. He defended a severe criticism which he had offered on some mediocre work by the remark, "J'aime mieux être incivil qu'ennuyé." With his dislike of sycophants and bores, it is not surprising that during his life he had more enemies than friends.

Euler now took the matter up and showed that for the wave equation the general solution is

$$u = f(x + ct) + g(x - ct)$$

where f and g are arbitrary functions.

Wave equation

As initiated by Taylor, the wave equation

$$\frac{\partial^2 u(x,t)}{\partial x^2} = \frac{1}{v^2}\frac{\partial^2 u(x,t)}{\partial t^2}$$

is this equation for free vibrations of a string. It is one dimensional, with the dimension x.

Taylor determined that the wave equation has sinusoidal solutions. A typical such solution would be

$$u(x,t) = \sin kx \cos \omega t$$

where k and ω are constants. Let us now verify that the above equation is indeed a solution. Its derivatives are

$$\frac{\partial u}{\partial x} = k \cos kx \cos \omega t$$

$$\frac{\partial^2 u}{\partial x^2} = -k^2 \sin kx \cos \omega t$$

$$\frac{\partial u}{\partial t} = -\omega \sin kx \sin \omega t$$

$$\frac{\partial^2 u}{\partial t^2} = -\omega^2 \sin kx \cos \omega t$$

If we substitute these second partial derivatives in the wave equation we obtain

$$-k^2 \sin kx \cos \omega t = \frac{1}{v^2}(-\omega^2 \sin kx \cos \omega t)$$

which gives the dispersion equation

$$k^2 = \frac{\omega^2}{v^2} \quad \text{or} \quad k^2 = \left(\frac{\omega}{v}\right)^2$$

Therefore the wave equation has the solution

$$u(x,t) = \sin kx \cos \omega t$$

provided that

$$\omega = \pm kv$$

Combining the above two equations, we have

$$u(x,t) = \sin kx \cos kvt$$

A sinusoidal solution is also known as simple harmonic motion. The legendary Greek mathematician Pythagoras (ca 600 BC) was the first to consider a purely physical problem in which harmonic analysis made its appearance. Pythagoras studied the laws of musical harmony by .endpoints. This problem excited scientists for almost 2500 years before the mathematical turning point arrived when Taylor recognized that the vertical displacement u(x, t) of the vibrating string satisfies the wave equation.

After the work of Taylor, the problem of constructing the solution of the wave equation was attacked by some of the greatest mathematicians of all time, and in so doing, they paved the way for the theory of spectrum analysis. The constant v is a physical quantity characteristic of the material of the string. Jean Ronde d'Alembert showed that v represents the velocity of the traveling waves on the string.

The two dimensional wave equation is found by adding the spatial dimension y; that is,

$$\frac{\partial^2 u}{\partial x^2} + \frac{\partial^2 u}{\partial y^2} = \frac{1}{v^2}\frac{\partial^2 u}{\partial t^2}$$

The three dimensional wave equation if found by adding the spatial dimension z as well.

$$\frac{\partial^2 u}{\partial x^2} + \frac{\partial^2 u}{\partial y^2} + \frac{\partial^2 u}{\partial z^2} = \frac{1}{v^2}\frac{\partial^2 u}{\partial t^2}$$

The time variable t, the space variables x, y, z, the velocity variable v, and the wave amplitude u are all continuous. For digital processing all of these quantities must be transformed into discrete series.

Traveling waves and standing waves

A traveling wave travels through a region without resulting in a net displacement of the material located in that region. For example, when a thrown rock causes ripples in a pond, those ripples pass and then the pond looks as it did before. A wave can transfer energy from one point to another without transferring material between the two points.

There are some basic definitions needed to describe a wave. A wavelength is the length between two consecutive similar points on a wave (peak to peak, valley to valley, etc...). The amplitude is maximum displacement of the wave, or the maximum height of the wave above the equilibrium position. The period is the time it takes for the wave to complete one wavelength. The frequency is how often a wavelength passes by a given point, which is also equal to one divided by the period.

Two waves with the same frequency, wavelength and amplitude traveling in opposite directions will interfere and produce a standing wave or stationary wave. The equations

$$u_1(x,t) = A \sin(kx - \omega t)$$
$$u_2(x,t) = A \sin(kx + \omega t)$$

represent sinusoidal waves going in opposite directions where:

A is the amplitude of each wave,
$\omega = 2\pi f$ is the angular frequency measured in radians per second
$k = 2\pi/\lambda$ is the angular wave number measured in radians per meter

The equation for the standing wave is given by

$$u = u_1 + u_2 - A\sin(kx - \omega t) + A\sin(kx + \omega t)$$

By use of trigonometric identities, the equation for the standing wave reduces to

$$u = 2A \sin(kx) \cos(\omega t)$$

The standing wave oscillates in time, but it also has the spatial dependence $\sin(kx)$ that is stationary. The locations

$$x = 0, \frac{\lambda}{2}, \lambda, \frac{3\lambda}{2}, \cdots$$

are called the nodes. The amplitude at a node is always zero. The locations

$$x = \frac{\lambda}{4}, \frac{3\lambda}{4}, \frac{5\lambda}{4}, \ldots$$

are called the anti-nodes. The amplitude (in magnitude) at an anti-node is maximum. The distance between two anti-nodes is λ/2.

Wavefronts and ray paths

In the simple case where the medium is uniform everywhere, the ray paths are straight lines. This leads to the principle of rectilinear propagation. According to this principle, the energy in a uniform medium always travels in straight lines. Thus, in a uniform medium, a bit of energy travels in a straight line from its locus at the source, or on any particular wavefront, to its associated locus on any succeeding wavefront. It is to be emphasized, however, that only in a homogeneous medium are all the ray paths straight lines.

This type of wave propagation may be illustrated by a disturbance originating at a point in a medium which is perfectly homogeneous. The disturbance propagates as a spherical wavefronts and radial ray paths from a point source in a uniform medium. The source itself could, for instance, be a shot of dynamite. Since the medium is uniform, the wave proceeds outward in all directions from the source, and its successive positions will assume the form of expanding concentric spheres with the source as center. As a consequence of uniformity, the velocity of the wave will be constant everywhere in the medium. The successive positions of the wave after equal intervals of elapsed time will, accordingly, be equally spaced spheres; i.e., spheres whose radii differ by equal increments. Each wavefront, which represents the locus of the disturbance after certain time interval, is a spherical arc. If the total elapsed time is t and the velocity is v, then the radius of the sphere is equal to vt.

Each radius represents a raypath associated with a spherical wavefronts. A raypath may be given physical significance by thinking of it as the path over which the energy of the disturbance travels from the source S to the spherical wavefront. It is evident in this simple case that the ray paths are everywhere at right angles to the wavefronts. It is true in general for isotropic media that the ray paths are always orthogonal to the wavefronts regardless of the wavefront shape.

For our present purpose, we define the leading-edge wavefront as the most forward position of the advancing region of disturbance at any particular instant of time. Behind the leading-edge wavefront, the medium has been disturbed. Ahead of the leading-edge wavefront, the medium is undisturbed. It is merely the physical effect caused by the original disturbance being propagated away from its source as a consequence of the natural elastic behavior of the medium.

Perhaps the simplest experiment illustrating a similar type of wave propagation in two dimensions, and the one with which we are most familiar, is that of tossing a stone into a pond of water. The initial water disturbance is set up at the point where the stone strikes the surface, and the appearance of the resultant circular expanding waves is a matter of common experience.

One-dimensional traveling waves

Suppose that a disturbance u travels in the positive x-direction with a constant positive velocity v. The specific nature of the disturbance is at the moment unimportant. Since the disturbance is moving, it must be a function of both position and time and can therefore be written as $u(x, t)$. The shape of the disturbance at any instant, say $t = 0$, can be found by holding time constant at that value. In this case,

$$u(x, t)]_{t=0} = u(x, 0) = f(x)$$

represents the shape or profile of the wave at that time. Once we choose the shape $f(x)$, we can substitute $x - vt$ for x in $f(x)$ and thus obtain

$$u(x, t) = f(x - vt)$$

The resulting expression $u(x, t)$ describes a wave traveling in the positive x-direction and having the desired profile. If we check the form of the above equation by examining u after an increase in time of Δt and a corresponding increase of $v\Delta t$ in x, we find

$$f[(x + v\Delta t) - v(t + \Delta t)] = f(x - vt)$$

which shows that the profile is unaltered.

Similarly, if the wave was traveling in the negative x-direction, i.e., to the left, we have

$$u(x,t) = f(x + vt)$$

where as before we assume that the quantity vc is a positive number.

For v a positive quantity, we may combine both directions of travel into the single equation

$$u(x,t) = f(x \pm vt)$$

with the negative sign indicating propagation in the positive x-direction, and the positive sign indicating propagation in the negative x-direction.

We may conclude, therefore, that regardless of the shape of the wave, the variables x and t must appear in the function as a unit, i.e., as a single variable in the form $(x - vt)$. This equation is often expressed equivalently as some function of $(t - x/v)$, since

$$f(x - vt) = F\left(-\frac{x - vt}{v}\right) - F\left(t - \frac{x}{v}\right)$$

The one-dimensional wave equation is

$$\frac{\partial^2 u(x,t)}{\partial x^2} = \frac{1}{v^2} \frac{\partial^2 u(x,t)}{\partial t^2}$$

Because it is a linear partial differential equation, it follows that if two different wave functions u_1 and u_2 are each separate solutions, then $u_1 + u_2$ is also a solution. Accordingly, the wave equation is most generally satisfied by a wave function having the form, called d'Alembert's formula,

$$u(x,t) = A_1 f(x + vt) + A_2 g(x - vt)$$

where A_1 and A_2 are constants. This is clearly a sum of two waves traveling in opposite directions along the x-axis with the same speed but not necessarily the same profile. The superposition principle is inherent in this equation, and the physical system is said to be linear.

Many physical systems involve the simultaneous propagation of two or more waveforms. Examples of this situation are especially common in acoustics. The sounds which we hear represent a complicated combination of various wave motions, resulting in some overall pattern. Within normal limits, the following basic assumption holds:

The resultant of two or more individual waves is simply the sum of the individual waves.

In the context of the wave equation, this assumption is a purely mathematical consequence. Usually, however, it becomes a physical question. Is the displacement produced by two disturbances, acting together, equal to the superposition of the displacements as they would be observed to occur separately? The answer to this question may be yes or no, according to whether or not the physical assumptions are satisfied which make the wave equation valid. Always the mathematical consequences of an equation must be separated from the question as to the physical conditions required in order to make the equation valid in the first place. At the moment, we are addressing ourselves to the purely mathematical properties implied by the wave equation; the physical applicability of the results is not involved at this point.

The behavior of a single pulse traveling in one direction is easily visualized. But what happens when one pulse moves from right to left at the same time that another pulse moves from left to right? As we know, the two pulses superimpose when they meet. The best way to understand this phenomenon is by means of a motion picture. What happens when two pulses travelling in opposite directions are started on the same string at the same time? The pulses approach each other as if each had the string to itself. As they cross each other, the two pulses combine to form complicated shapes. But, after having crossed, they again assume their original shapes and travel along the string as if nothing had happened. If we perform this experiment over and over again with different pulses, we always get the same result.

The fact that two pulses pass through each other without either being altered is a fundamental property of waves. However, if we throw two tennis balls in opposite directions, and if they hit each other, then their motion is violently changed. The crossing of waves is thus a very different process than the crossing of streams of balls made out of solid matter.

Let us now take a closer look at the superposition that occurs when two pulses cross each other. Often the shape of the combined pulse does not resemble the shape of either of the two original pulses. However, we can see the relationship, as follows. We visualize each of the original pulses at the position it would occupy as if alone, and then add the displacements of both original pulses to get the resultant pulse. In other words, the principle of superposition says that the resultant displacement of any point on the string at any instant is equal to the sum of the displacements that would have been produced by the two pulses independently. As a matter of fact, the principle works for more than two pulses; the resulting displacement for any number of pulses is the sum of the displacements for the individual pulses.

The *principle of superposition* can be summarized as follows. To find the form of the total wave displacement at any time, add at each point the displacements belonging to each pulse that is passing through the medium. This simple addition gives the actual displacement of the medium.

Let us now apply the superposition principle to the case of two equal symmetrical pulses but with opposite polarity. The two pulses are assumed to have exactly the same shape and size and each is symmetrical. Suppose that the one that displaces the string upward is the one that travels to the right. The pulse that displaces the string downward travels to the left. The addition of equal displacements upward (plus) and downward (minus) leave us with a net displacement of zero. As the pulses pass each other, there is a time point at which the whole string appears undisplaced. It represents the situation of complete cancelation. How is this situation different from the case of a string at rest? In such a case, (that is, when the string carries no wave motion), all the various pieces of the string stand still at all times. On the other hand, when two symmetric equal and opposite waves are passing, there is only one instant when the string is passing through its rest position, but at that instant the string is moving. All of the wave energy at that instant is in the form of the kinetic energy of the moving string.

Sinusoidal waves

Let us now examine the simplest wave form where the profile is a sine or cosine curve. These are known as sinusoidal waves, or by the older

Chapter 4. Wave motion

terminology as simple harmonic waves. Because, by Fourier's theorem, any periodic can be synthesized by a superposition of sinusoidal waves, it is important that we know their properties.

Choose as the profile the simple function

$$u(x,t)]_{t=0} = u(x,0) = A \sin kx$$

where k is a positive constant known as the *propagation number* and kx is in radians. The sine varies from $+1$ to -1 so that the maximum value of $u(x,0)$ is A. This maximum disturbance is known as the *amplitude* of the wave. In order to transform the above equation into a progressive wave traveling at speed v in the positive x direction, we need to replace x by $(x - vt)$, in which case

$$u(x,t) = A \sin k(x - vt) = A \sin(kx - kvt)$$

This is a solution of the wave equation. If we hold either x or t fixed, the result is a sinusoidal disturbance. Thus the wave is periodic in both space and time.

The spatial period is known as the *wavelength* and is denoted by λ. An increase or decrease in x by the amount λ leaves u unaltered; i.e.,

$$u(x,t) = u(x \pm \lambda, t)$$

which is

$$A \sin(kx - kvt) = A \sin[k(x \pm \lambda) - kvt]$$

In the case of a sinusoidal wave, this equation is equivalent to altering the argument of the sine function by $\pm 2\pi$. which is

$$A \sin(kx - kvt) = A \sin(kx - kvt \pm 2\pi)$$

Let us equate the right hand sides of the above two equations. We obtain

$$A \sin[k(x \pm \lambda) - kvt] = A \sin(kx - kvt \pm 2\pi)$$

which gives

$$k(x \pm \lambda) - kvt = kx - kvt \pm 2\pi$$

Clearing terms, the above equation gives

$$\pm k\lambda = \pm 2\pi$$

and so

$$|k\lambda| = 2\pi$$

If both k and λ are taken as positive numbers, we obtain the fundamental equation

$$k = \frac{2\pi}{\lambda}$$

The wavelength λ is the distance between successive crests of the spatial sinusoidal wave and is commonly expressed in units of meters.

In a completely analogous fashion, we can examine the *temporal period* T, usually called just the *period*. It is the amount of time it takes for one complete wave to pass a stationary observer. In this case, it is the repetitive behavior of the wave in time which is of interest, so that

$$u(x,t) = u(x, t \pm T)$$

In this case, we find that

$$A\sin[kx - kv(t \pm T)] = A\sin(kx - kvt \pm 2\pi)$$

This equation gives

$$\pm kvT = \pm 2\pi$$

and so

$$|kvT| = 2\pi$$

If both k and T are taken as positive numbers, we obtain the fundamental equation

$$kvT = 2\pi$$

which is

$$kvT = 2\pi$$

Thus we have

$$T = \frac{2\pi}{kv}$$

Because

$$k = \frac{2\pi}{\lambda}$$

we have

$$T = \frac{\lambda}{v}$$

Chapter 4. Wave motion

The period T is the units of time per cycle. The inverse of the period is the frequency f, which is cycles per unit of time. Thus,

$$f = \frac{1}{T}$$

Frequency f is commonly expressed in Hertz (Hz), which stands for cycles per second.

The velocity v is commonly expressed in units of "meters per second". The two quantities which are often used in the literature of wave motion are the angular frequency

$$\omega = \frac{2\pi}{T}$$

commonly expressed in radians per second and the angular wavenumber (given above)

$$k = \frac{2\pi}{\lambda}$$

commonly expressed in radians per meter. We immediately see that the *angular wavenumber* is the same thing as what we previously called the *propagation number*. The wavelength, wavenumber, period, and frequency all describe aspects of the repetitive nature of a wave in space and time. These concepts can be applied to waves which are not sinusoidal as long as each wave is made up of a regularly repeating pattern (i.e., periodic waves.)

If we divide the wave velocity v (expressed in meters per second) by the wavelength λ (expressed in meters), the length term (meters) cancels out, and the result must be expressed as "something per second." As we know, we call this result frequency f, because it states how frequently the new wave crests pass a given point. Frequency f is expressed in cycles per second, because it tells us how many wave troughs and crests (i.e., how many cycles) pass the given point in one second. More specifically, we can say that wavelength λ is expressed as meters per cycle. Then we have

$$\frac{\lambda}{v} = \frac{\text{meters per cycle}}{\text{meters per second}} = \frac{\text{cycles}}{\text{second}} = f$$

Thus we have

$$f\lambda = \text{frequency} \times \text{wavelength} = \text{velocity} = v$$

In a medium of constant velocity, we see immediately that a wave with a short wavelength has a high frequency, and one with a long wavelength a low frequency.

For example, suppose that we have a machine that generates the same pulse shape, one after the other, at equal intervals T. In doing this, the wave generator repeats its motion once every interval T, the period of the motion. If the motion repeats every 0.01 second, then the frequency f is 100 Hz (i.e., 100 cycles per second). Let us now concentrate on some spatial point. The pulses produced by the generator move toward this point, and they pass the point with the same frequency that they leave the source. The frequency of the wave motion is therefore also 100 Hz, and the time between passage of successive pulses is also 0.01 second. Furthermore, as the waves move, the spatial distance between any two adjacent pulses is always the same and is the wavelength λ. Because the pulses are separated by a distance λ and each pulse moves over this distance in time T, it follows the velocity of propagation $v = \lambda/T$. Using the relation

$$f = \frac{1}{T}$$

we find that $v = f\lambda$, or that the velocity of propagation of a periodic wave is the product of the frequency and the wavelength. This is an important relationship, and in particular it applies to sinusoidal waves.

Now we come to an application of the formula $v = f\lambda$. Instead of watching a periodic wave continuously, we look at it through a shutter that is closed most of the time and opens periodically for short time intervals. Such an instrument is the stroboscope. The first time the shutter opens, we see the wave pattern in a certain position. During the time duration that the shutter is closed, all the pulses move a distance equal to their velocity times that time duration. As we look through the shutter as it is periodically opening and closing, the pattern will usually appear to move. However, if the period of the shutter is the same as the period of the wave motion, then, during the time the shutter is closed, each pulse moves up to the position of the pulse just ahead of it.

Consequently, we see the same pattern each time the shutter opens. In other words, we see a stationary pattern from which it is easy to measure the wavelength λ. In addition, as we have said, the period of the shutter is equal to the period T of the wave, and thus we can obtain it by simply counting the number of times that the shutter is opened each second; i.e., by measuring the frequency of the shutter. Now we have both f and T for the wave, so we can make use of the formula

$$v = f\lambda$$

in order to determine the velocity of the wave.

Phase velocity

Examine the sinusoidal wave function

$$u(x,t) = A\sin(kx - \omega t)$$

The argument of the sine function is known as the phase ϕ of the wave, so that

$$\phi = kx - \omega t$$

At $t = x = 0$, we have

$$u(0,0) = 0$$

which is certainly a special case. We can more generally write a sinusoidal wave as

$$u(x,t) = A\sin(kx - \omega t + \varepsilon)$$

where ε is the initial phase. The phase of this sinusoidal wave is

$$\phi(x,t) = kx - \omega t + \varepsilon$$

The partial derivative of ϕ with respect to t holding x constant is the rate of change of phase with time, and is equal to the negative frequency; that is,

$$\frac{\partial \phi}{\partial t} = -\omega$$

Similarly, the rate of change of phase with distance x holding t constant is the wavenumber; that is,

$$\frac{\partial \phi}{\partial x} = k$$

The condition of constant phase is expressed as

$$\phi(x,t) = kx - \omega t + \varepsilon = \text{constant}$$

Taking differentials, we have

$$d\phi(x,t) = k\, dx - \omega\, dt = 0$$

Solving the equation, we have (for ϕ = constant)

$$\left[\frac{\partial x}{\partial t}\right]_\phi = \frac{\omega}{k} = -\frac{\frac{\partial \phi}{\partial t}}{\frac{\partial \phi}{\partial x}}$$

The term on the left of the above equation represents the velocity of propagation of the condition of constant phase. Choose any point on the profile; e. g., the crest of the wave. As the wave moves through space, the displacement u of the point remains constant. Since the only variable in the sinusoidal wave function is the phase, it too must be constant. That is, the phase is fixed at such a value as to yield the constant u corresponding to the chosen point. The point moves along with the profile at the velocity v and so does the condition of constant phase as well.

Velocity is equal to frequency times wavelength, i.e., $v = f\lambda$. Because $\omega = 2\pi f$ and $\lambda = 2\pi/k$, it follows that $v = \omega/k$. Therefore

$$\left[\frac{\partial x}{\partial t}\right]_\phi = \frac{\omega}{k} = v$$

Thus, the speed at which the profile moves is the wave velocity v or, more specifically, as the phase velocity. The phase velocity carries a positive sign when the wave moves in the direction of increasing x and a negative one in the direction of decreasing x.

Consider the idea of the propagation of constant phase and how it relates to any one of the sinusoidal waves, say

$$u(x,t) = A \sin k(x \pm vt)$$

Let us assume that the velocity v is positive and let us choose the negative sign, so the condition of constant phase is

$$\phi = k(x - vt) = \text{constant}$$

This equation says that as t increases, x increases. In other words, the constant-phase condition moves in the increasing x-direction. Now let us again assume c is positive but choose the positive sign in the equation, so that the condition of constant phase is

$$\phi = k(x + vt) = \text{constant}$$

This equation says that as t increases, x decreases. . In other words, the constant-phase condition moves in the decreasing x-direction.

Further points

Up to this point, we have pictured a wave as a periodic function, that is, as a wave train that involves a whole succession of crests and troughs of the same shape, but this is not at all necessary. In fact, in seismic exploration, innumerable situations occur in which we picture a single isolated pulse of a seismic disturbance as propagating from one place to another. Pulses of this sort can be set up by taking a stretched string and producing in it a local deformation, by pulling one end and then holding it still. At any instant only a limited region of the string is disturbed, and the regions before and after are quiescent. The pulse travels in this way until it reaches the far end of the string, at which point it is reflected. However, as long as the pulse continues uninterrupted, it preserves the same shape.

How can we relate the behavior of a pulse to what we have already learned of sinusoidal waves? The answer is obtained by Fourier analysis, but in terms of the Fourier integral instead of Fourier series.

In order for a geophysicist to carry out valid seismic interpretations, he must have a good understanding of the mechanism by which elastic waves are propagated through various materials. He should know how the transmission is affected by the many types of rock formations and structures through which the waves pass. The clearer his understanding of these physical processes and the better his background knowledge of the geology, the sounder will be his interpretation. If the recorded events consist of distinct wave forms, reflected or refracted from essentially horizontal and plane subsurface strata, their processing and interpretation usually presents no great difficulty. However, if the travel paths have been influenced by irregular subsurface structural features, the seismic recordings become very complicated. Under these circumstances, the deleterious effects of scattering, refraction, focusing, and other disturbing influences may be so severe as to render the records extremely difficult to process and interpret.

In order to gain the most comprehensive knowledge of the details of wave propagation in the earth, it is necessary to undertake a study of the theory of elasticity. This very extensive subject deals with the elastic properties of many types of media and attempts to explain the manner in which solids, such as rocks, are deformed under the stress of applied forces. It further seeks to explain all possible types and modes of reactions, both static and dynamic, which result when a solid is acted upon by external agencies. It describes in detail how motion, momentum, and energy, initiated by a local disturbance, are propagated to other parts of an extended mass. This theory provides the most satisfactory exposition of the complex processes of elastic wave propagation in solids. The general theory has been a subject for investigation by mathematical physicists for many generations, and, as a consequence, it has attained a high degree of completeness. However, the theory involves much difficult mathematics and, in the end process, the actual solutions that can be obtained must be done with a computer.

Fortunately, there is a simpler approach which we can follow to explain many of the phenomena connected with seismic wave propagation. Seismic waves, in many respects, are similar to light waves. During the past several centuries, physicists and mathematicians have built up a very extensive body of principles, based almost entirely on geometrical reasoning, to explain what is known about light. This doctrine deals exhaustively with the ideas of ray paths and wavefronts and is known as *geometric optics*. It is readily possible to adapt many of these principles to explain seismic waves as well as acoustic waves. These fields of study are known as *geometric seismics* and *geometric acoustics*. Although this simpler approach does not suffice to explain everything in which we are interested as geophysicists, it is quite adequate to explain many important phenomena. Insofar as it is capable of dealing with problems in seismic wave propagation, it agrees with the dynamical elastic theory.

Chapter 4. Wave motion

In the more complex areas of oil exploration, the effectiveness of the seismic method will depend to a large extent upon the ability of geophysicists to interpret the more difficult data. This in turn depends largely upon a thorough understanding of the mechanism of seismic wave propagation. The principles of geometrical seismics provide a very good insight into many of the essential physical processes which constitute the source of the original data.

The geometrical theory of wave propagation deals with wavefronts and ray paths. In a uniform medium, the ray paths are straight lines at right angles to the wavefronts. A simple example of wave propagation is that of water waves. From a point disturbance on the surface of a pond, for instance, circular waves proceed outward in all directions. The concentric circles are the wavefronts and the radii of the circles are the ray paths. The geometrical theory of wave propagation does not deal in any way with such specific rock properties as density, compressibility, and rigidity. In fact, it is incapable of dealing with them. Nor can it deal quantitatively with such important matters as the motion and momentum of the portions of a body traversed by a wave. Likewise, the energy content and the shape of the wave, or the distinction between longitudinal and transverse waves cannot be treated. Of all the elastic properties of a rock, only one of them, namely the velocity with which it transmits elastic waves, is used in the geometrical theory.

In some rocks, especially the crystalline types, the velocity at any point may be different depending on the direction of propagation of the wave through the point. For instance, the velocity of propagation in a vertical direction may differ from that in a horizontal direction at the same locality. When this is the case, the rock is called anisotropic. An example of an anisotropic rock is Icelandic spar. Unless otherwise stated, we will deal only with rocks which are isotropic; i.e., those in which the velocity of propagation of seismic waves is independent of the direction.

There is another type of inhomogeneity with which we will often be concerned. The rocks may be isotropic, everywhere but the absolute velocity may very from place to place. Normally, the velocity in sedimentary rocks increases with depth. This is called vertical velocity variation. The velocity may also change in a horizontal direction. This is called horizontal or lateral velocity variation. When these variations are encountered, methods must be devised to properly take them into account.

When the velocity is constant everywhere, the medium is said to be uniform or homogeneous. A great deal of our work will be with media of this type.

Exercises

1. What is an oscillation? Answer: An oscillation occurs when a body repeatedly travels between two points on opposite sides of equilibrium.

2. What two things do all oscillations have in common? Answer: (1) The oscillation is centered about an equilibrium position. (2) The oscillation is periodic.

3. When does simple harmonic motion (SHM) occur? It occurs when a linear restoring force is applied to restore an object to its equilibrium position.

4. Relate uniform circular motion is simple harmonic motion. Answer: When projected onto one dimension, uniform circular motion is simple harmonic motion. A particle traveling at angular velocity ω in uniform circular motion matches the motion of a simple harmonic oscillator oscillating with angular frequency ω.

5. Describe the terms used in SHM. Answer. SHM is sinusoidal and characterized by: Displacement, amplitude, angular frequency, time, phase φ by means of the equation $x(t) = A\cos(\omega t + \varphi)$. The displacement x of a particle during an oscillation is the object's position from equilibrium during an oscillation. The amplitude A of motion is the maximum displacement from equilibrium. The period T of an oscillation is the time it takes to complete one oscillation. Frequency f is the number of oscillations per second and the unit of frequency is the Hertz. Hertz stands for *cycles per second*. Frequency is inversely proportional to period. Angular frequency ω is the *frequency of angular oscillation* and is related to frequency f and period T by $\omega = 2\pi f = 2\pi/T$.

Chapter 4. Wave motion

Angular frequency for SHM is equivalent to angular velocity for a particle in uniform circular motion.

6. By differentiating displacement, show that that the velocity and acceleration of an object moving with simple harmonic motion $x(t) = A\cos(\omega t + \varphi)$ are $v(t) = -\omega A \sin(\omega t + \varphi)$ and $a(t) = -\omega^2 x(t)$

7. At $t = 0$, the displacement $x(0)$ of an object in simple harmonic motion is -8.50 cm. The velocity $v(0)$ of the object is -0.920 m/s, and its acceleration $a(0)$ is $+47.0$ m/s^2. What is the angular frequency ω, the phase constant φ and amplitude A?

8. What makes SHM different from some other forms of repeated motion? Answer. It is the way that the force (and therefore the acceleration) changes during each cycle. In SHM, there is a restoring force which always acts toward the equilibrium position, no matter where the object is. The further the object moves from equilibrium, the greater the force becomes. At the equilibrium position the resultant force is zero. If an object is moving with SHM the following two conditions must apply: 1. The acceleration a of the object is directly proportional to the displacement x from the equilibrium position. 2. The acceleration is always directed toward the equilibrium position.

9. Standing waves can be regarded as linear superpositions of traveling waves, of equal amplitude and wavenumber, propagating in opposite directions. In other words, standing waves are not fundamentally different to traveling waves. Write the standing wave

$$u(x,t) = \sin kx \cos kvt$$

as a superposition of two traveling waves propagating in opposite directions.

10. Write the traveling wave) $u(x,t) = A\sin(kx - \omega t)$ as a superposition of two standing waves.

11. During exercise, a person's heart beats 60 times in 20 seconds. At rest, it beats 36 times in 30 seconds. Calculate the frequency and period. Answer. (1). During exercise. 60 beats in 20 seconds gives $60/20$ beats per second $= 3.0$ Hz and $T = 1/f = 1/3 = 0.33$ s. (2). At

rest, 36 beats in 30 seconds gives 36/30 beats per second = 1.2 Hz and $T = 1/f = 1/1.2 = 0.83$ s

12. As the frequency of a wave decreases, the period: (a) decreases (b) increases (c) remains the same

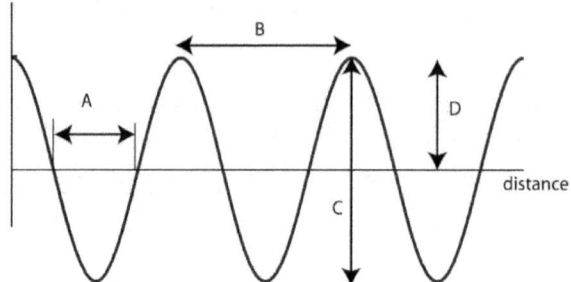

13. In the above figure, the wavelength is given by which letter? The amplitude is given by which letter? How many complete wavelengths are represented?

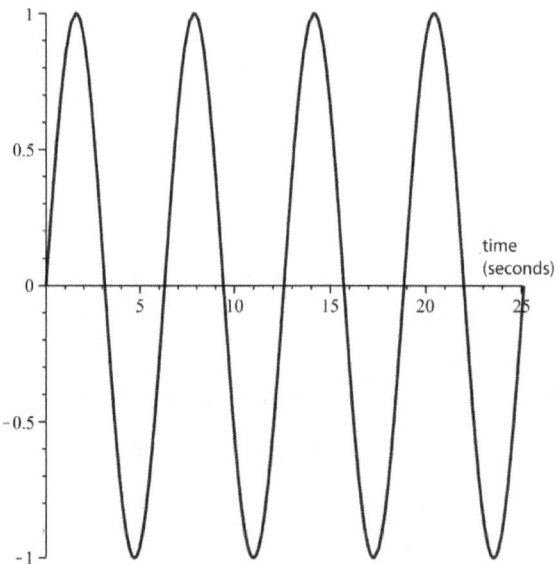

14. In the above figure, how many periods are shown? What if you wanted to fit more periods into the 25 seconds shown, would you shorten or lengthen the period? If you fit more periods into the 25 seconds shown, would you be increasing or decreasing the frequency of the wave? It is important to understand how waves propagate (travel).

Whether a seismic wave, a sound waves or a light wave, the speed of a wave can always be defined as:

$$\text{speed} = \text{wavelength times frequency}$$

In the above figure, what is the period for the wave? Recall that

$$\text{frequency} = 1/\text{period}$$

What is the frequency?

MIT Whirlwind computer

On August 6, 1952, the MIT Industrial Liaison Office held a meeting entitled "A Conference on the Generalized Harmonic Analysis of Seismograms." Those attending included representatives from thirteen major oil companies as well as from Texas Instruments and United Geophysical. Two theoretical and numerical Reports were distributed to the participants. The results of everything done to date were presented at the meeting. Enders Robinson demonstrated by numerical examples that the Whirlwind digital computer did every step in the deconvolution of digital seismic records.

The following passage is from the 1953 film on MIT Project Whirlwind, *Making Electrons Count*. "The film which you are about to see first shows a few examples of the types of problems in which computers can be useful, and then describes the efforts of a typical user in programming a problem for Whirlwind. Whirlwind has been involved in more than a hundred such computations problems, originating in many different departments of MIT. Take the Geology Department, for example. Seismic methods of prospecting for oil may seem a little strange to the onlooker. A charge is exploded at one point, and the sound, reflected from various underground layers of rock, is recorded at a number of other points. A great deal of information about underground formations can be determined from these sound patterns, but only after long and tedious computations have been performed on them."

Chapter 5. Hamilton's equations and seismic modeling

Gauss: Theory attracts practice as the magnet attracts iron.

Wave-particle duality

A ray is a line drawn in space that corresponds to the direction of flow of radiant energy. As such, it is a mathematical device rather than a physical entity. In practice, one can produce very narrow beams or pencils (as for example a laser beam) and we might imagine a ray to be the unattainable limit on the narrowness of such a beam. In an isotropic medium, i. e., one whose properties are the same in all directions, ray trajectories are orthogonal to wavefronts. That is to say, they are lines normal to the wavefronts at every point of intersection. In such a medium a ray is parallel to the propagation vector k. However, this is not true in anisotropic materials.

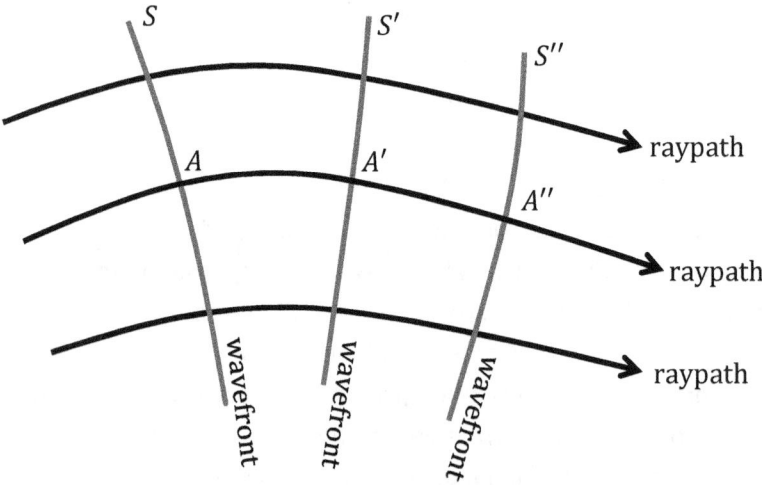

Fig. 1. Raypaths, wavefronts, and corresponding points

Within homogeneous isotropic materials, rays are straight lines since by symmetry they cannot bend in any preferred direction, there being none. Moreover, because the speed of propagation is identical in all

Chapter 5. Hamilton's equations and seismic modeling

directions within the given medium, the spatial separation between two wavefronts, measured along rays, must be the same everywhere. Points where a single ray intersects a set of wavefronts are called *corresponding points*; as for example A, A' and A'' in Fig. 1.

Evidently, the separation in time between any two corresponding points on any two sequential wavefronts is identical. In other words, if a wavefront S transforms into a wavefront S' after a time t', the distance between corresponding points on any and all rays will be traversed in the same time t'. This is true even if the wavefronts pass from one homogeneous isotropic medium into another. It means that each point on S can be viewed as one following the path of a ray arriving at S' in a time t'. Light can often be thought of as rays.

A water wave as seen on a pond appears as a set of moving wavefronts. If a stone is thrown into a pond, the crests (wavefronts) form a pattern of concentric circles. The energy of the disturbance travels outward radially from the center. That is, the energy is propagated along rays at right angles to the wavefront. If we carefully watch the wave motion, we observe that the longest waves appear at the outside of the expanding pattern of concentric circles. As we watch the progress of one of these outside crests, we will suddenly see them disappear. However, this is not an illusion, for the next crest coming from behind also disappears. More and more crests keep coming from behind and disappear at the outside edge. On the inside of the ring, new crests keep appearing from the now calmed central water. The reason for this phenomenon is as follows. The wave packet represented by the concentric rings moves outward at the group velocity. The group velocity is the velocity at which the energy of the disturbance propagates outward. The crests, however, move at the phase velocity.

From the physics it can be shown that the phase velocity is greater than the group velocity. Thus, the crest of each wave moves faster than the wave packet, and the crests move forward with respect to the concentric pattern. Once a crest reaches the outside of the pattern, it cannot go on any further, as no energy has yet reached there, and thus the crest can only disappear. In Hamilton's wave-particle dualism, the wave packet represents a particle moving along a ray, and the group velocity is the velocity of the particle. The crest of each wave moves faster than the wave packet as a whole, and the crests move forward with respect to the concentric pattern. Once a crest reaches the outside of the pattern, it cannot go any further, since no energy has yet arrived there, and so the crest simply vanishes.

In many physical problems we can find the dispersion relation for the wave equation governing the wave motion. In symbols let us write such a dispersion relation as the mathematical equation

$$\omega = \omega(r, k)$$

where ω is the frequency, r is the position vector, and k is the wavenumber vector. We may think of $\omega(r, k)$ as a surface in 6-dimensional space. For a fixed value of r, the subsurface is the wavenumber surface in 3D space k. The gradient to this 3D surface is the group velocity

$$v_g = \nabla_k \omega = \frac{\partial \omega}{\partial k}$$

For a fixed value of k, the subsurface is a surface in 3D space r. The gradient to this 3D surface is

$$\nabla_r \omega = \frac{\partial \omega}{\partial r}$$

These two gradients play an interesting role in the Hamilton theory.

Example 1. Consider a two dimensional vertically stratified medium:

$$r = (x, z), \quad k = (k_x, k_z), \quad v(x, z) = v(z)$$

By definition, the velocity depends only upon the depth coordinate z, and not the horizontal coordinate x. The dispersion relation is

$$\omega = \omega(r, k) = v(z)(k_x^2 + k_z^2)^{1/2}$$

so ω is a function of the form $\omega(k_x, k_y, z)$. The group velocity vector is

$$v_g = \left(\frac{\partial \omega}{\partial k_x}, \frac{\partial \omega}{\partial k_z}\right) = \left(\frac{k_x v}{k}, \frac{k_z v}{k}\right) = [\alpha v(z), \gamma v(z)]$$

where the wavenumber k is $(k_x^2 + k_z^2)^{1/2}$ and where $\alpha = \sin \theta$ and $\gamma = \cos \theta$ are the direction cosines of the k vector. See Fig. 2.

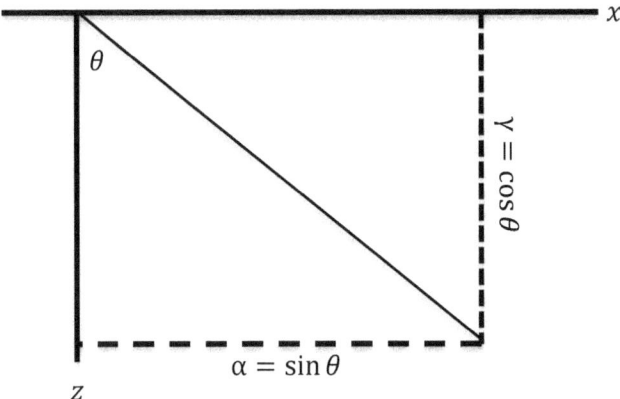

Fig. 2. The direction cosines

Here θ is the angle that $\mathbf{k}$ makes with the z-axis. Because the group velocity is independent of frequency ω there is no dispersion. The other gradient vector is the vector

$$\frac{\partial \omega}{\partial \mathbf{r}} = \left(\frac{\partial \omega}{\partial x}, \frac{\partial \omega}{\partial z}\right) = \left[0, v'(z)k\right]$$

This vector has a zero component in the x direction, so that variations can only take place in the z direction, as we would expect for a vertically stratified medium.

Let us now develop the Hamilton wave-particle duality. We start our treatment with the dispersion equation $\omega = \omega(\mathbf{r}, \mathbf{k})$. The basic assumption we make for the medium is that frequency and wavenumber vary slowly. That is, we assume (1) that the frequency ω does not change greatly in one oscillation period, and (2) that the wavenumber vector $\mathbf{k}$ does not change much in magnitude and direction over a distance of one wavelength.

We recall that a *plane wave* in a homogeneous medium can be written in the form

$$\exp(i\varphi) = \exp i(\omega t - \mathbf{k} \cdot \mathbf{r})$$

where ω and $\mathbf{k}$ are constant. The quantity

$$\varphi = \omega t - \mathbf{k} \cdot \mathbf{r}$$

is called the phase. For such a plane wave we see that

$$d\varphi = \omega\, dt - \mathbf{k} \cdot d\mathbf{r} = \left[\omega - \mathbf{k} \cdot \frac{d\mathbf{r}}{dt}\right] dt$$

so we can write the phase as

$$\varphi = \int d\varphi = \int \left[\omega - \mathbf{k} \cdot \frac{d\mathbf{r}}{dt}\right] dt$$

We now make use of our basic assumption of slow variation of ω and $\mathbf{k}$. That is, ω and $\mathbf{k}$ are approximately constant with respect to period and wavelength respectively. Thus, we assume that for our slowly-varying medium the same equation holds, namely, the equation

$$\varphi = \int_{t_0}^{t_1} \left[\left(\omega(\mathbf{r}, \mathbf{k})\right) - \mathbf{k} \cdot \frac{d\mathbf{r}}{dt}\right] dt$$

In physical terms the quantity within square brackets in the above equation is called the Hamiltonian and the phase φ is called the action.

According to *Hamilton's principle of least action*, the required solution for the ray path is found by minimizing the action, i.e., by minimizing the above integral. The magnitude of the integral depends upon the mathematical function chosen for the ray path $\mathbf{r}(t)$ as a function of t. In order to examine the relationship between the action (i.e., phase) φ and the function $\mathbf{r}(t)$, it is convenient to calculate the change in φ for the transition from some arbitrary function $\mathbf{r}(t)$ to another infinitely close, but also arbitrary, function $\mathbf{r}_1(t)$.

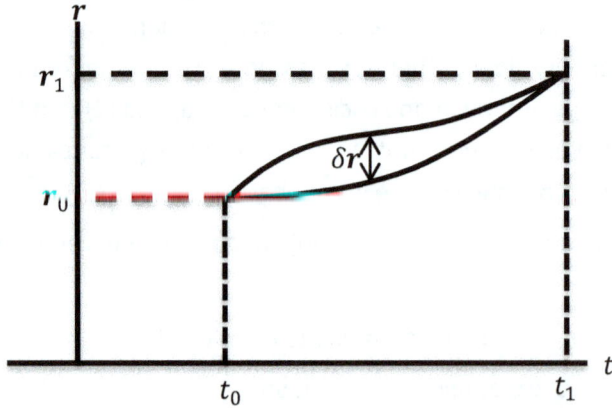

Fig. 3. Two possible raypaths with the same starting point and with the same endpoint

Chapter 5. Hamilton's equations and seismic modeling

Fig. 3 shows two such conceivable paths, where time t is plotted along the abscissa and r is schematically plotted on the ordinate. We assume that all such paths pass through the same points $r_0 = r(t_0)$ and $r_1 = r(t_1)$ at the initial time t_0 and final time t_1, respectively. The vertical distance between two such paths at some instant of time is called the variation of r and denoted by δr. At the end points t_0 and t_1, of course, $\delta r = 0$ because all paths coincide at these points by assumption. The reason that the symbol δ is used is that we want to make clear the difference between the variation δ and the differential d. The differential is taken for the same path at various instants of time, whereas the variation is taken for the same instant of time between different paths.

The variation in the action (phase) is given by

$$\delta \varphi = \delta \int_{t_0}^{t_1} \left[\omega(r, k) - k \cdot \frac{dr}{dt} \right] dt$$

which is

$$\delta \varphi = \int_{t_0}^{t_1} \left[\frac{\partial \omega}{\partial k} \cdot \delta k + \frac{\partial \omega}{\partial r} \cdot \delta r - (\delta k) \cdot \frac{dr}{dt} - k \cdot \left(\delta \frac{dr}{dt} \right) \right] dt \quad (1)$$

In the above equation, the two partial derivatives of ω are gradients. The gradient of ω with respect to k is

$$\nabla_k \omega = \frac{\partial \omega}{\partial k} = \left(\frac{\partial \omega}{\partial k_x}, \frac{\partial \omega}{\partial k_y}, \frac{\partial \omega}{\partial k_z} \right)$$

and the gradient of ω with respect to r is

$$\nabla_k \omega = \frac{\partial \omega}{\partial r} = \left(\frac{\partial \omega}{\partial x}, \frac{\partial \omega}{\partial y}, \frac{\partial \omega}{\partial z} \right)$$

The last term inside parentheses in equation (1) above can be integrated by parts. Doing so we get

$$\int_{t_0}^{t_1} \left[-k \cdot \left(\delta \frac{dr}{dt} \right) \right] dt = -k \cdot \delta r \Big]_{t_0}^{t_1} + \int_{t_0}^{t_1} \left(\frac{dk}{dt} \cdot \delta r \right) dt$$

Because all curves $r(t)$ pass through the same endpoints (as previously stated), the integrated part becomes zero; i.e.,

$$-\mathbf{k} \cdot \delta \mathbf{r}]_{t_0}^{t_1} = 0$$

which gives

$$\int_{t_0}^{t_1} \left[-\mathbf{k} \cdot \left(\delta \frac{d\mathbf{r}}{dt} \right) \right] dt = \int_{t_0}^{t_1} \left(\frac{d\mathbf{k}}{dt} \cdot \delta \mathbf{r} \right) dt$$

Thus, equation (1) for the variation in the phase becomes

$$\delta \varphi = \int_{t_0}^{t_1} \left[\frac{\partial \omega}{\partial \mathbf{k}} \cdot \delta \mathbf{k} + \frac{\partial \omega}{\partial \mathbf{r}} \cdot \delta \mathbf{r} - (\delta \mathbf{k}) \cdot \frac{d\mathbf{r}}{dt} + \frac{d\mathbf{k}}{dt} \cdot \delta \mathbf{r} \right] dt$$

which is

$$\delta \varphi = \int_{t_0}^{t_1} \left[\delta \mathbf{k} \cdot \left(-\frac{d\mathbf{r}}{dt} + \frac{\partial \omega}{\partial \mathbf{k}} \right) + \delta \mathbf{r} \cdot \left(\frac{d\mathbf{k}}{dt} + \frac{\partial \omega}{\partial \mathbf{r}} \right) \right] dt \quad (2)$$

The independent variables are now $\mathbf{k}$ and $\mathbf{r}$. The variations $\delta \mathbf{k}$ and $\delta \mathbf{r}$ are completely arbitrary. Thus for $\delta \varphi$ to be zero, each of the following two equations must be satisfied:

$$\frac{d\mathbf{r}}{dt} = \frac{\partial \omega}{\partial \mathbf{k}} \quad \text{and} \quad \frac{d\mathbf{k}}{dt} = -\frac{\partial \omega}{\partial \mathbf{r}} \quad (3)$$

These two equations are called *Hamilton's equations*.

We will now show that the frequency ω stays constant along a ray path. The rate of change of frequency

$$\omega = \omega(\mathbf{r}, \mathbf{k})$$

with time for arbitrary rates of change of $\mathbf{r}, \mathbf{k}$ is the total derivative

$$\frac{d\omega}{dt} = \frac{\partial \omega}{\partial \mathbf{k}} \cdot \frac{d\mathbf{k}}{dt} + \frac{\partial \omega}{\partial \mathbf{r}} \cdot \frac{d\mathbf{r}}{dt}$$

Substituting Hamilton's equations (3) into the right-hand side of the above equation, we obtain zero; that is,

$$\frac{d\omega}{dt} = \frac{d\mathbf{r}}{dt} \cdot \frac{d\mathbf{k}}{dt} + \left(-\frac{d\mathbf{k}}{dt} \right) \cdot \frac{d\mathbf{r}}{dt} = 0$$

Thus for the rates of changes along the ray, as given by Hamilton's equations, the derivative $d\omega/dt$ is zero. Hence, frequency ω remains constant along a ray.

Let us now state Hamilton's duality between waves and particles. Let the coordinate vector $\mathbf{r}$ represent the coordinates of both wave and

Chapter 5. Hamilton's equations and seismic modeling

particle. Then the wavenumber vector $\mathbf{k}$ of the wave corresponds to the momentum vector of the particle. The frequency ω of the wave corresponds to the energy of the particle. As we have seen, the group velocity is the velocity with which the wave packet travels. The group velocity is

$$v_g = \frac{\partial \omega}{\partial \mathbf{k}}$$

The velocity with which the particle travels along the ray is

$$\frac{d\mathbf{r}}{dt}$$

The first Hamilton equation is

$$\frac{\partial \omega}{\partial \mathbf{k}} = \frac{d\mathbf{r}}{dt}$$

Thus the first Hamilton equation states that the group velocity $\partial \omega / \partial \mathbf{k}$ with which the wave packet travels is the same as the velocity $d\mathbf{r}/dt$ with which the particle travels along the ray. This result is the essence of the wave particle duality. Although the wave particle duality was established by Sir William Hamilton in 1825, the physical significance of this duality was not understood until 1924 when Prince Louis de Broglie suggested that an electron has a dual character (that is, an electron is a particle with laws of motion that are wave-like in character). This wave-particle dualism for the electron matched the wave-particle dualism for the photon as worked out by Arthur Compton in 1923.

Ray tracing

If we solve Hamilton's equations, we can trace the path of the ray. This procedure is called *ray tracing*. Let us give some examples.

Example 2. In a *constant velocity medium* we have

$$\omega(\mathbf{r}, \mathbf{k}) = k v = \left(k_x^2 + k_z^2\right)^{1/2} v$$

The component form of Hamilton's equations give

$$\frac{dx}{dt} = \frac{\partial \omega}{\partial k_x} = \frac{k_x v}{k} \quad \text{and} \quad \frac{dz}{dt} = \frac{\partial \omega}{\partial k_z} = \frac{k_z v}{k} \quad (4)$$

and

$$\frac{dk_x}{dt} = -\frac{\partial \omega}{\partial x} = 0 \quad \text{and} \quad \frac{dk_z}{dt} = -\frac{\partial \omega}{\partial z} = 0 \quad (5)$$

Equations (5) say respectively that k_x and k_z are constant. The magnitude of the vector

$$\mathbf{k} = (k_x, k_z)$$

is

$$k = |\mathbf{k}| = \left(k_x^2 + k_z^2\right)^{1/2}$$

Thus k is also constant. Integrating equations (4) we obtain

$$x = \frac{k_x}{k} vt + x_0 \quad \text{and} \quad z = \frac{k_z}{k} vt + z_0$$

where x_0 and z_0 are the values of x and z respectively at $t = 0$. Thus, the equation of the ray is

$$(x - x_0, z - z_0) = (k_x, k_z)\frac{vt}{k}$$

Thus the ray is in the direction of the constant propagation vector (k_x, k_z). Since k is the magnitude of the propagation vector, we see that we have a plane wave moving with constant velocity v.

Example 3. Let us now consider a *stratified medium* with the dispersion relation

$$\omega = kv(z) = \left(k_x^2 + k_z^2\right)^{1/2} v(z)$$

In this case Hamilton's equations are the four equations. The *first two* are

$$\frac{dx}{dt} = \frac{\partial \omega}{\partial k_x} = \frac{k_x v(z)}{k} \quad \text{and} \quad \frac{dz}{dt} = \frac{\partial \omega}{\partial k_z} = \frac{k_z v(z)}{k}$$

and the *second two* are

$$\frac{dk_x}{dt} = -\frac{\partial \omega}{\partial x} = 0 \quad \text{and} \quad \frac{dk_z}{dt} = -\frac{\partial \omega}{\partial z} = kv'(z)$$

Dividing dx/dt by dz/dt as given in the first two, we get the direction of the ray, which is

Chapter 5. Hamilton's equations and seismic modeling

$$\frac{dx}{dz} = \frac{\frac{dx}{dt}}{\frac{dz}{dt}} = \frac{\frac{\partial \omega}{\partial k_x}}{\frac{\partial \omega}{\partial k_z}} = \frac{\frac{k_x v(z)}{k}}{\frac{k_z v(z)}{k}} = \frac{k_x}{k_z} \quad (6)$$

Because

$$\mathbf{k} = (k_x, k_z)$$

the ratio

$$\frac{k_x}{k_z}$$

gives the direction of the wavenumber vector $\mathbf{k}$. As we have just shown,

$$\frac{dx}{dz} = \frac{k_x}{k_z}$$

Therefore the rays are in the same direction as the wavenumber vector.

The first equation in the second two gives $k_x = $ constant. We now use our previous general result that frequency ω is constant along a ray. Thus, k_x/ω is constant along a ray in a stratified medium. If we call this constant p_x, then we have

$$p_x = \frac{k_x}{\omega} = \frac{k_x}{kv(z)} = \text{constant}$$

along a ray. We next define $\theta(z)$ as the angle between the ray and the vertical; that is,

$$\sin\theta = \frac{k_x}{k} \quad \text{and} \quad \cos\theta = \frac{k_z}{k}$$

Thus

$$p_x = \frac{k_x}{kv(z)} = \text{constant}$$

is seen to be Snell's law

$$p_x = \frac{\sin\theta}{v(z)} = \text{constant}$$

See Fig. 4.

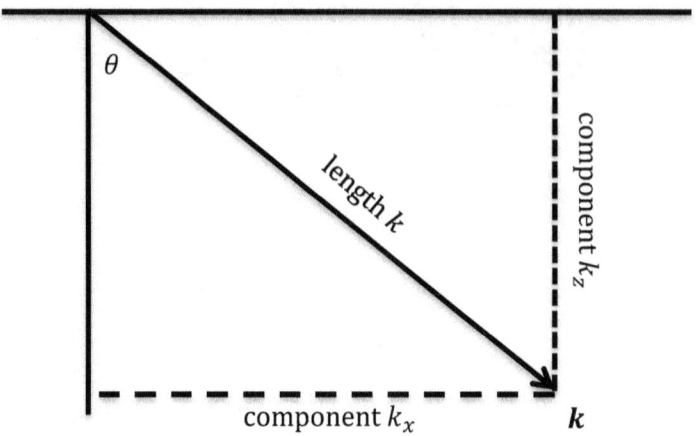

Fig. 4. Components of the wavenumber vector k

Using Snell's law, we have

$$\frac{k_x}{k} = \sin \theta(z) = p_x \, v(z)$$

$$\frac{k_z}{k} = \cos \theta(z) = [1 - p_x^2 \, v^2(z)]^{1/2}$$

We will now integrate the equation (6) which is

$$\frac{dx}{dz} = \frac{k_x}{k_z} = \frac{p_x \, v(z)}{[1 - p_x^2 \, v^2(z)]^{1/2}}$$

Doing so we obtain the so-called *horizontal distance equation*

$$x = \int_0^z \frac{p_x \, v(z)}{[1 - p_x^2 \, v^2(z)]^{1/2}} \, dz$$

We can also integrate the second Hamilton equation in the first two of the four, which is

$$\frac{dz}{dt} = \frac{k_z v(z)}{k} = [1 - p_x^2 \, v^2(z)]^{1/2} \, v(z)$$

Thus

$$\frac{dt}{dz} = \frac{1}{[1 - p_x^2 \, v^2(z)]^{1/2} \, v(z)}$$

From which we obtain the so-called *time equation*

Chapter 5. Hamilton's equations and seismic modeling

$$t = \int_0^z \frac{1}{[1 - p_x^2 \, v^2(z)]^{1/2} \, v(z)} \, dz$$

The time equation together with the horizontal distance equation) make up the so-called time-distance relationship for the ray. This time-distance relationship gives the so-called *time-distance curve* (as a function of the parameter p_x of waves originating at $z = 0$ and traveling to depth $z = $ constant. See Fig. 5.

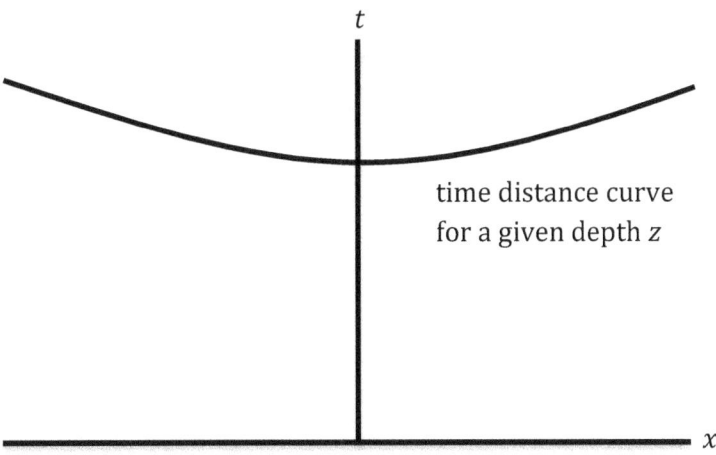

Fig. 5. The time-distance curve for a given depth

Each point on the time-distance curve is determined by a particular value of p_x. Since each value of p_x designates one ray, the time-distance curve summarizes information in all the rays reaching depth z.

Above we have derived the *horizontal distance equation*

$$x = \int_0^z \frac{p_x \, v(z)}{[1 - p_x^2 \, v^2(z)]^{1/2}} \, dz$$

Let us write the horizontal distance equation in another form. The first of the Hamilton equations in (4) is

$$\frac{dx}{dt} = \frac{k_x v(z)}{k} = v(z) \sin \theta(z) \quad \text{(along a ray)}$$

However, Snell's equation is $\sin \theta(z) = p_x \, v(z)$. Thus, the above equation give

$$\frac{dx}{dt} = p_x \, v^2(z) \quad \text{(along a ray)}$$

so the required *alternative form of the horizontal distance equation* is

$$x = \int_0^t p_x \, v^2(z) \, dt = p_x \int_0^t v^2(z) \, dt \qquad (7)$$

Remember that Hamilton's equations (4) and (5) apply for paths along a ray; thus, for example, $dx/dt = p_x \, v^2(z)$ along a ray. On the other hand, the time-distance curve applies to all rays (each characterized by a value of p_x) which extend to a given depth

$$z = \text{constant}$$

Now we want to show that

$$\frac{dx}{dt} = \frac{1}{p_x} \quad \text{(on the time distance curve)}$$

The geometric argument is shown in Fig. 6.

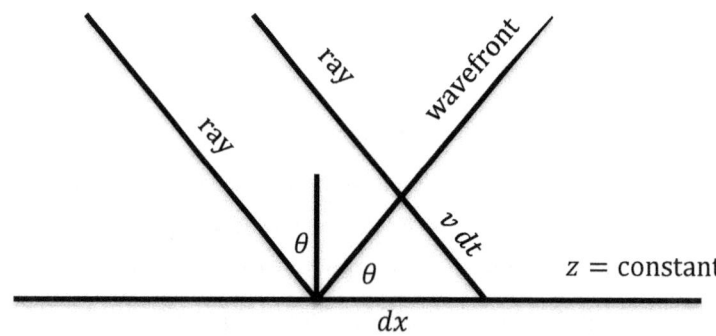

Fig. 6. Rays and the incident wavefront

From the diagram we see that

$$\frac{dx}{v \, dt} = \frac{1}{\sin \theta}$$

Using Snell's law $p_x = \sin \theta / v$, we obtain the required result; namely, $dx/dt = 1/p_x$. Thus, the slope of the time-distance curve is equal to Snell's parameter p_x; that is,

Chapter 5. Hamilton's equations and seismic modeling

$$\frac{dt}{dx} = p_x \quad \text{(on the time distance curve)}$$

The significance of the parameter p_x is realized by taking $z = 0$ in Snell's equation,

$$p_x = \frac{\sin \theta(0)}{v(0)}$$

The above equation shows that p_x is proportional to the sine of the incidence angle $\theta(0)$ of the ray at the surface $z = 0$.

In the case of constant velocity (i.e., $v(z) = V = $ constant), the time-distance curve is the hyperbola

$$V^2 t^2 - x^2 = z^2 = \text{constant} \tag{8}$$

See Fig. 7.

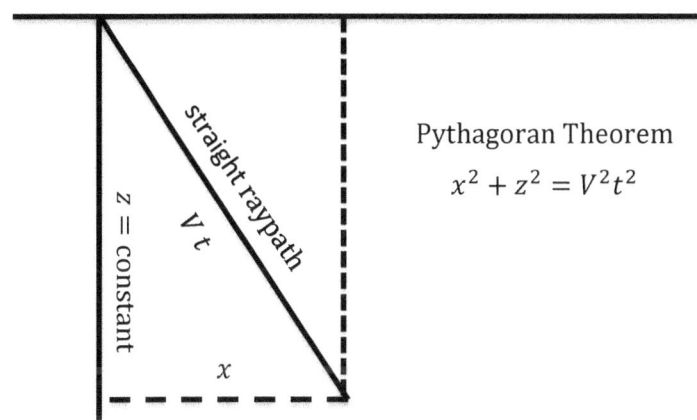

Fig. 7. The Pythagorean theorem

In the case of a stratified medium with velocity function $v(z)$, generally the time-distance curve will resemble a hyperbola in shape. Suppose a spike at time $t = 0$ propagates outward from a point source at $z = 0$. At some depth (or alternatively height) $z = $ constant, we measure the arrival time t of the spike as a function of the horizontal coordinate x. The result is the time-distance curve, such as the one shown in Figure 5.

Suppose we now fit the time-distance curve to the hyperbola (8), where the constant velocity $v(z) = V = $ constant is determined so as to yield the best fit. The question is: How does the fitted constant velocity V relate to the variable velocity function $v(z)$?

If we differentiate equation (8)) for the hyperbola, we obtain

$$2V^2 t\, dt - 2x\, dx = 0$$

so

$$\frac{dt}{dx} = \frac{x}{V^2 t} \quad \text{(on the hyperbola)}$$

However, on the time-distance curve we have

$$\frac{dt}{dx} = p_x$$

Thus, we set the two derivatives equal to obtain

$$p_x = \frac{x}{V^2 t}$$

Thus, the required value is

$$V^2 = \frac{x}{p_x t}$$

If we make use of equation (7), namely,

$$x = \int_0^t p_x v^2(z)\, dt = p_x \int_0^t v^2(z)\, dt$$

we obtain

$$V^2 = \frac{x}{p_x t} = \frac{1}{t} \int_0^t v^2(z)\, dt$$

The above equation says that V is the mean-square value of $v(z)$ along the ray path. Thus if we fit the time-distance curve by a hyperbola, the velocity V that we obtain is the RMS velocity given by

$$V = V_{\text{RMS}} = \left[\frac{1}{t}\int_0^t v^2(z)\, dt\right]^{1/2}$$

In words, we can say that if we need some average velocity to characterize the stratified medium to some depth z, an excellent choice is the root-mean-square velocity.

Chapter 5. Hamilton's equations and seismic modeling

Eikonal equation

In the general case for a spatially varying velocity function

$$v(\mathbf{r}) = v(x.y.z)$$

we can always solve Hamilton's equations numerically or graphically. The procedure is analogous to that employed in the mechanics of a particle moving in a three dimensional potential field for which the acceleration is a function of position.

In order to gain insight let us recast Hamilton's equations in a slightly different form. We assume an *isotropic medium*. Again we start from first principles, and proceed in an intuitive way. The *seismic traveltime field* $t(\mathbf{r})$ can be defined as the value of the traveltime from a convenient reference wave surface S_0 to an arbitrary point P with position vector $\mathbf{r}$. It is implicitly understood that P can be reached by a ray from S_0. See Fig. 8.

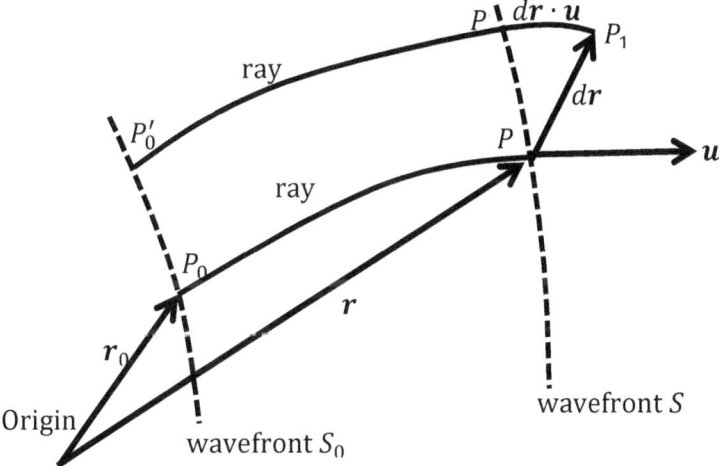

Fig. 8. Two wavefronts and two rays

Denote the point on S_0 at the foot of the ray by P_0 with position vector $\mathbf{r}_0$. Then the seismic traveltime field is

$$t(\mathbf{r}) = \int_{r_0}^{r} p \, ds$$

where

$$p(r) = \frac{1}{v(r)}$$

is the slowness and ds is an increment of path length along the given ray. It is understood, of course, that the path of integration is along the given ray.

Suppose that a seismic event at time zero is represented by the reference wavefront S_0. Then at another time, $t = $ constant, the seismic event will be represented by the wavefront S. Wavefront S consists of all points that can be reached in the time interval

$$t = \text{constant}$$

by rays starting on S_0. The equation defining S is the locus of positions r satisfying

$$t(r) = t = \text{constant}$$

Let P be the point where the ray from P_0 intersects wavefront S. Suppose now that we consider another point P_1, which is close to P. The distance from P to P_1 is dr. What is the corresponding change in the traveltime as r moves from P to P_1?

Let us first construct the ray that goes from the reference surface S_0 to P_1. We call this ray

$$P_0'\, P'\, P_1$$

where the intermediate point P' lies on the surface S. The distances $P_0' P'$ and $P_0 P_1$ are approximately the same, as they are distances between the two wave surfaces S and S_0. Thus the change in traveltime is due chiefly to the path length $|P'\, P_1|$.

The *unit tangent vector* is

$$u = (\alpha, \beta, \gamma) \quad \text{where} \quad \alpha^2 + \beta^2 + \gamma^2 = 1$$

The *unit tangent vector* at P is tangent to the ray $P_0 P$. The traveltime increment at P is

$$dt = p\,|P'\,P_1| = p\,u \cdot dr$$

From differential calculus, we know that in general dt is

$$dt = \frac{\partial t}{\partial x}dx + \frac{\partial t}{\partial y}dy + \frac{\partial t}{\partial z}dz = \nabla t \cdot dr$$

Chapter 5. Hamilton's equations and seismic modeling

It follows therefore that the gradient of the seismic traveltime function is

$$\nabla t = p\, u$$

or

$$\left(\frac{\partial t}{\partial x}, \frac{\partial t}{\partial y}, \frac{\partial t}{\partial z}\right) = (p\alpha, p\beta, p\gamma) \qquad (9)$$

The ray from P_0 cuts the wavefront S at P. This equation says that, at P, the gradient of the wavefront is equal to the slowness p times the unit tangent vector to the ray. Therefore the gradient and the unit tangent vector both point in the same direction. Since the gradient is orthogonal to the wavefront, and the tangent is along the ray, it follows that ray is orthogonal to the wavefront. Since $|u| = 1$, the absolute value of equation (9) gives a result known as the *eikonal equation*

$$|\nabla t| = p$$

The eikonal equation, in square form, is

$$\nabla t \cdot \nabla t = p^2$$

or

$$\left(\frac{\partial t}{\partial x}\right)^2 + \left(\frac{\partial t}{\partial y}\right)^2 + \left(\frac{\partial t}{\partial z}\right)^2 = \frac{1}{v^2(x.y.z)}$$

Example 4. Let us consider a two-dimensional stratified medium, with horizontal coordinate x and depth z, and with velocity function $v(z)$. The surface of the earth is the line $z = 0$. The standard time-distance curve which depicts a seismic event received on the surface of the earth is, in the present notation,

$$t(x, z = 0) = t(x, 0)$$

Often we simply use the notation $t(x)$ for $t(x, 0)$. Let the emergence angle of the ray be θ. Then the unit tangent vector to the ray is

$$u = (\alpha,\ \beta) = (\sin\theta,\ \cos\theta)$$

where $\alpha = \sin\theta$ and $\beta = \cos\theta$ are the direction cosines of u. The eikonal equation is (since $p = 1/v$)

$$\left(\frac{\partial t}{\partial x}, \frac{\partial t}{\partial z}\right) = \left(\frac{\sin\theta}{v}, \frac{\cos\theta}{v}\right)$$

The first component

$$\frac{\partial t}{\partial x} = \frac{\sin\theta(z)}{v(z)} = p_x$$

states that the derivative of the time-distance curve is equal to the Snell parameter p_x.

Ray equations

The seismic ray at any given point follows the direction of the gradient of the traveltime field $t(r)$. As before, let $\boldsymbol{u}$ be the unit tangent vector along the ray. The ray in general will follow a curved path, and $\boldsymbol{u}$ will be the tangent to this curved raypath. We now want to derive an equation that will tell us how $\boldsymbol{u}$ changes along the curved raypath. We write the vector $\boldsymbol{u}$ as

$$\boldsymbol{u} = (\alpha, \beta, \gamma) \quad \text{where } (\alpha^2 + \beta^2 + \gamma^2)^{1/2} = 1$$

In this section, the position vector r always represents a point on a specific raypath, and not any arbitrary point in space. As time increases, r traces out the particular raypath in question. Let us now find how a general function of position $f(r)$ will change along the raypath curve. For a general displacement dr, we have from calculus that

$$df = \frac{\partial f}{\partial x} dx + \frac{\partial f}{\partial y} dy + \frac{\partial f}{\partial z} dz = \nabla f \cdot d\boldsymbol{r}$$

Because, by assumption, $d\boldsymbol{r}$ is along the ray-path curve, it has length ds. We can thus write

$$d\boldsymbol{r} = \boldsymbol{u}\, ds \tag{10}$$

where $\boldsymbol{u}$ is the unit tangent vector to the raypath curve. We thus obtain

$$\boldsymbol{u} = (\nabla f \cdot \boldsymbol{u})\, ds$$

so the directional derivative of f along the raypath curve is

$$\frac{df}{ds} = \boldsymbol{u} \cdot \nabla f$$

Since the curve is a raypath curve, the eikonal equation

Chapter 5. Hamilton's equations and seismic modeling

$$\nabla t = p\, \mathbf{u}$$

holds, so

$$\frac{df}{ds} = \frac{1}{p}\nabla t \cdot \nabla f = \frac{1}{p}(\nabla t \cdot \nabla f)$$

We now make a special choice for f, namely,

$$f = \frac{\alpha}{v} = p\alpha$$

From the eikonal equation

$$\left(\frac{\partial t}{\partial x}, \frac{\partial t}{\partial y}, \frac{\partial t}{\partial z}\right) = (p\alpha, p\beta, p\gamma)$$

we have

$$f = p\alpha = \frac{\partial t}{\partial x}$$

Thus the above equation

$$\frac{df}{ds} = \frac{1}{p}(\nabla t \cdot \nabla f)$$

gives

$$\frac{d(p\alpha)}{ds} = \frac{d}{ds}\frac{\partial t}{\partial x} = \frac{1}{p}\left(\nabla t \cdot \nabla \frac{\partial t}{\partial x}\right)$$

Now comes some algebra. The dot produced in the expression on the right can be written out in full, and the terms collected to give

$$\frac{d(p\alpha)}{ds} = \frac{1}{2p}\frac{\partial}{\partial x}\left[\left(\frac{\partial t}{\partial x}\right)^2 + \left(\frac{\partial t}{\partial y}\right)^2 + \left(\frac{\partial t}{\partial z}\right)^2\right]$$

We know the square of the eikonal equation is

$$\left(\frac{\partial t}{\partial x}\right)^2 + \left(\frac{\partial t}{\partial y}\right)^2 + \left(\frac{\partial t}{\partial z}\right)^2 = p^2$$

Thus

$$\frac{d(p\alpha)}{ds} = \frac{1}{2p}\frac{\partial(p^2)}{\partial x} = \frac{\partial p}{\partial x}$$

Similar equations hold for the y and z components. We are thus led to the vector equation

$$\frac{d(p\boldsymbol{u})}{ds} = \frac{\partial p}{\partial \boldsymbol{r}}$$

This equation, together with equation (10), which we write as

$$\frac{d\boldsymbol{r}}{ds} = \boldsymbol{u}$$

are called the *ray equations*. In summary, the *two ray equations* are

$$\frac{d\boldsymbol{r}}{ds} = \boldsymbol{u} \quad \text{and} \quad \frac{d(p\boldsymbol{u})}{ds} = \frac{\partial p}{\partial \boldsymbol{r}} \tag{11}$$

From Hamilton's equations to ray equations

The frequency ω, wavenumber k, and velocity v are related by

$$\omega = k\,v$$

An increment ds of distance s along a raypath and an increment dt of traveltime t are related by

$$ds = v\,dt$$

The first Hamilton equation is

$$\frac{d\boldsymbol{r}}{dt} = \frac{\partial \omega}{\partial \boldsymbol{k}}$$

Because

$$\frac{d\boldsymbol{r}}{dt} = \frac{d\boldsymbol{r}}{ds}\frac{ds}{dt} = v\frac{d\boldsymbol{r}}{ds}$$

It follows that the *first Hamilton equation* becomes

$$v\frac{d\boldsymbol{r}}{ds} = \frac{\partial(kv)}{\partial \boldsymbol{k}} \tag{12}$$

The propagation vector $\boldsymbol{k}$ is

$$\boldsymbol{k} = (k_x, k_y, k_z)$$

Its magnitude k is

$$k = |\boldsymbol{k}| = \left(k_x^2 + k_y^2 + k_z^2\right)^{1/2}$$

The velocity $v(x, y, z)$ is a function of x, y, z, whereas k is considered as the function

$$k = \left(k_x^2 + k_y^2 + k_z^2\right)^{1/2}$$

Chapter 5. Hamilton's equations and seismic modeling

Thus equation (12) becomes

$$v \frac{d\mathbf{r}}{ds} = v \frac{\partial k}{\partial \mathbf{k}}$$

Canceling v from each side of the above equation, we obtain

$$\frac{d\mathbf{r}}{ds} = \frac{\partial k}{\partial \mathbf{k}} \qquad (13)$$

The left hand side of (13) is the vector

$$\frac{d\mathbf{r}}{ds} = \mathbf{u} = (\alpha, \ \beta, \ \gamma) \quad \text{where } \alpha^2 + \beta^2 + \gamma^2$$

which we recognize as the unit tangent vector to the raypath curve.

The right hand side of (13) is the vector

$$\frac{\partial k}{\partial \mathbf{k}} = \left(\frac{\partial k}{\partial k_x}, \ \frac{\partial k}{\partial k_y}, \ \frac{\partial k}{\partial k_z} \right) = \left(\frac{k_x}{k}, \ \frac{k_y}{k}, \ \frac{k_z}{k} \right) = \frac{\mathbf{k}}{k}$$

which we recognize as the gradient of k with respect to the variables k_x, k_y, k_z. The vector $\mathbf{k}/k$ as the unit vector in the $\mathbf{k}$ direction.

Thus the *first Hamilton equation* gives

$$\frac{d\mathbf{r}}{ds} = \mathbf{u}$$

This equation is the *first ray equation*. We have thus shown *that the first Hamilton equation is the same as the first ray equation.*

Next consider the second Hamilton equation

$$\frac{d\mathbf{k}}{dt} = -\frac{\partial \omega}{\partial \mathbf{r}}$$

We can write this equation as

$$v \frac{d\mathbf{k}}{ds} = -\frac{\partial (kv)}{\partial \mathbf{r}}$$

which is

$$v \frac{d\mathbf{k}}{ds} = -k \frac{\partial v}{\partial \mathbf{r}} \qquad (14)$$

We now use the equation

$$\mathbf{k} = k\mathbf{u} = \frac{\omega}{v} \mathbf{u} = \omega p \mathbf{u}$$

where $p = 1/v$ is the slowness. The second Hamilton equation becomes with the aid of equation (14)

$$p^{-1}\frac{d(\omega p \mathbf{u})}{ds} = -\omega p \frac{\partial(p^{-1})}{\partial \mathbf{r}}$$

which is

$$p^{-1}\omega \frac{d(p\mathbf{u})}{ds} = \omega p^{-1}\frac{\partial p}{\partial \mathbf{r}}$$

or

$$\frac{d(p\mathbf{u})}{ds} = \frac{\partial p}{\partial \mathbf{r}}$$

This equation is the *second ray equation*. We have thus shown *that the second Hamilton equation is the same as the second ray equation.*

Numerical ray tracing

Let us now consider the general case in which we have a spatially varying velocity function $v(\mathbf{r}) = v(x, y, z)$. This velocity function represents a velocity field. For a fixed constant v, the equation $v(\mathbf{r}) = v$ specifies those positions $\mathbf{r}$ which have this fixed value of v. The locus of such positions makes up an *isovelocity surface*. The gradient

$$\nabla v(\mathbf{r}) = \left(\frac{\partial v}{\partial x}, \frac{\partial v}{\partial y}, \frac{\partial v}{\partial x}\right)$$

is normal to the isovelocity surface and points in the direction of the greatest increase in velocity. Similarly, the equation $p(\mathbf{r}) = p$ for a fixed value of slowness p specifies an *isoslowness surface*. The gradient

$$\nabla p(\mathbf{r}) = \left(\frac{\partial p}{\partial x}, \frac{\partial p}{\partial y}, \frac{\partial p}{\partial x}\right)$$

is normal to the isoslowness surface and points in the direction of greatest increase in slowness. The isovelocity and isoslowness surfaces coincide, and

$$\nabla v(\mathbf{r}) = -p^{-2}\nabla p(\mathbf{r})$$

so the respective gradients point in opposite direction, as we would expect.

Chapter 5. Hamilton's equations and seismic modeling

A seismic ray makes its way through the slowness field. As the wavefront progresses in time, the raypath is bent according to the slowness field. For example, suppose we have a stratified earth in which the slowness decreases with depth. A vertical raypath will not bend, as it is pulled equally in all lateral directions. However, a nonvertical ray will drag on its slow side, so it will curve away from the vertical and bend toward the horizontal.

This is the case of a diving wave, whose raypath eventually curves enough to reach the earth's surface again. Certainly, the slowness field, together with the initial direction of the ray, determines the entire raypath. Except in special cases, however, we must determine such raypaths numerically.

Assume we know the slowness function $p(r)$ and that we know the ray direction u_1 at point r_1. See Fig. 9.

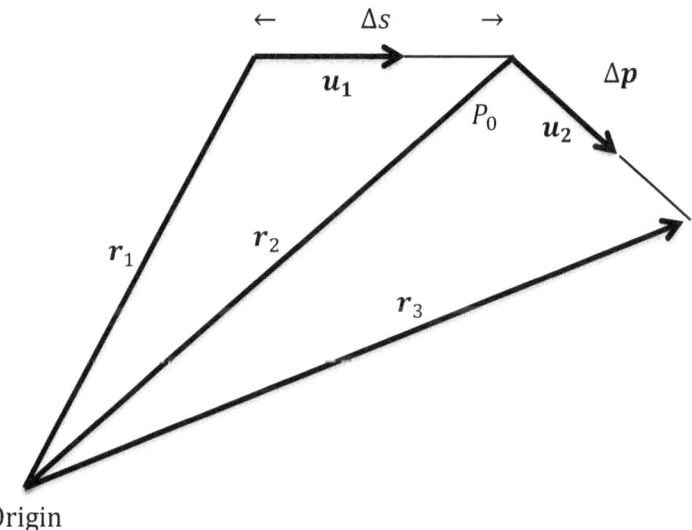

Fig. 9. Algorithm for finding the ray direction

We now want to give an algorithm for finding the ray direction u_2 at point r_2. We choose a small, but finite change in path length Δs. Then we use the *first ray equation*, which we recall is

$$\frac{dr}{dx} = u$$

to compute the change

$$\Delta r = r_2 - r_1$$

The required approximation is

$$\Delta r = u_1 \, \Delta s$$

or

$$r_2 = r_1 + \Delta r = r_1 + u_1 \, \Delta s$$

We have thus found the first desired quantity r_2.

We next use the *second ray equation*, which we recall is

$$\frac{d(p\,u)}{ds} = \nabla p$$

in the form

$$d(p\,u) = \nabla p \, ds$$

The required approximation is

$$\Delta(p\,u) = \nabla p \, \Delta s$$

or

$$p(r_2)\,u_2 - p(r_1)\,u_1 = \nabla p \, \Delta s$$

For accuracy, ∇p may be evaluated by differentiating the known function $p(r)$ midway between r_1 and r_2. Thus the desired u_2 is given as

$$u_2 = \frac{p(r_1)}{p(r_2)} u_1 + \frac{\Delta s}{p(r_2)} \nabla p$$

Note that the vector u_1 is pulled in the direction of ∇p in forming u_2. This is, the ray drags on the slow side, and so is bent in the direction of increasing slowness. The special case of no bending occurs when u_1 and ∇p are parallel. As we have seen, a vertical wave in a stratified medium is an example of such a special case.

We have thus found how to advance the wave along the ray by an incremental raypath distance. We can repeat the algorithm to advance the wave by any desired distance.

Chapter 6. Predictive deconvolution

Gauss: Mathematical discoveries, like springtime violets in the woods, have their season which no human can hasten or retard.

Introduction

A least-squares digital filter converts, in the least squares sense, an input signal into a desired output signal. The design of such a filter requires two entities. One is the autocorrelation of the input signal and the other is the crosscorrelation of the desired output signal and the input signal. The least-squares prediction filter is a special case of the least-squares filter. For this case, the desired output is a time-advance version of the input signal. All that is required for the design of a least-squares prediction filter is the autocorrelation of the input signal.

Least-squares filtering, as a mathematical process, does not require any model for the internal parts of the signals involved. It is this aspect that makes least-squares filtering so applicable. The design of a least-squares filter is the same whatever the signal involved. Least-squares filters are used for economic time series, such as the such a stock market closing prices, meteorological records, medical recordings such a EKG data, and seismic traces.

Deconvolution and least-squares filtering are two different but related things. Least-squares filtering does not require no model. Deconvolution requires a model that describes the internal structure of the digital signal. More exactly, deconvolution requires a convolutional model. With a convolutional model, least-squares filters can be used to carry out the required deconvolution process.

The ensemble of the reflection coefficients makes up the reflectivity series. The seismic trace is the response of the reflectivity series to the wavelet. That is the trace is a superposition of the individual wavelets. This linear process is called the principle of superposition. It is achieved computationally by convolving the wavelet with the reflectivity series. To identify closely spaced reflecting boundaries, the wavelet must be removed from the trace to obtain the reflectivity series. This removal process is just the opposite of the convolutional process used to obtain the response of the reflectivity series to the wavelet. The reverse process appropriately is called deconvolution.

The convolutional model states that a seismic trace is additively composed of many overlapping seismic wavelets which arrive as time progresses. The trace is the convolution of the wavelet and a reflectivity series. This convolutional model yields a representation of a seismic trace at any moment in terms of its own observable past history plus an unpredictable, random-like innovation. The standard assumption is that each wavelet has the same minimum-phase shape and that the arrival times and strengths are given by a reflectivity series of uncorrelated random variables. The spiking deconvolution operator can be computed from the autocorrelation of the trace. This operator can be used to deconvolve the trace, that is, the spiking operator removes the wavelet from the trace, thereby yielding the reflectivity series. In addition, the spiking operator can be inverted to yield the basic wavelet shape. Both the spiking operator and the wavelet are minimum-phase.

Other least-squares filters can be derived. For example, shaping filters can be computed by least squares. A shaping filter shapes the given wavelet into another wavelet that may be more appropriate.

Deconvolution improves the temporal resolution of a seismic section by compressing the seismic wavelets. For example, the reverberatory (ringy) character of the marine record without deconvolution limits resolution considerably. Deconvolution can remove these reverberations. A spiking deconvolution filter is a prediction error filter with prediction distance equal to one time unit. The method of gap deconvolution makes use of least-squares prediction error filters with prediction distances greater than unity. In the case of water reverberations, prediction error filters with prediction distance greater than one are often used. When model driven deconvolution is applicable, prediction error filters can be use to deconvolve a trace generated by a mixed phase wavelet. The resulting deconvolved trace is the convolution of the white reflectivity series convolved with a short mixed-phase orphan wavelet.

Chapter 6. Predictive deconvolution

It is commonly believed that gap deconvolution represents a more general approach to deconvolution than spike deconvolution. It is often stated that gap deconvolution allows one to control the length of the desired output wavelet, and hence to specify the desired degree of resolution. Moreover a commonly occurring statement is that gap deconvolution attempts to preserve the main lobes of the wavelet and requires no assumption as to the phase of the wavelet. The purpose of this chapter is to relate gap deconvolution to spike deconvolution. It turns out that gap deconvolution relies just as much on the minimum-phase assumption for the wavelet as does spike deconvolution. Not in any way is gap deconvolution a more generalized approach to deconvolution than spike deconvolution. Gap deconvolution is merely a filtered version of spike deconvolution, the filter being the head of th minimum-phase wavelet.

Least-squares prediction and filtering

Various types of filtering operations can be applied to digital signals. One operation is that of extrapolation; in other words, of prediction. The prediction, of course, does not in general give a precise continuation of the digital signal, for, if there is new information to come, this completely precludes an exact estimate of the future. Digital signals that have a statistical nature are subject to a statistical prediction. This means that the continuation of a series is estimated which minimizes the statistically determinable quantity known as the mean square error. This consideration also applies to whatever other operations that can be performed on signals.

Another important operation performed upon signals is that of purification. Very often the signal is observable only after it has been in some way corrupted or altered by mixture, additively or not, with other signals. It is of importance to ascertain as nearly as possible in an appropriate sense, that is, in a statistical sense, what the data would have been like without the contamination of the other signals. Filtering may be the complete problem to be addressed; or it may be combined with a prediction problem, which means that it is desired to know what the uncontaminated signal will do in the future. Other cases might allow a certain amount of leeway in time and the problem then is to determine what the uncontaminated signal series would have been in a certain past epoch.

While the purification problem is clearly distinguishable from the prediction problem, mixed problems involving elements of both are of great importance. The pure purification problem is that in which noise is to be eliminated in real time without an attendant time delay. In many practical circuit problems, a uniform delay is not undesirable provided that the delay is not excessive. In many case, good filter performance depends on the introduction of a delay. If the delay is negative, the filter becomes a filtering predictor, which is often a useful instrument.

The geophysicists that process seismic data are among some of the world's most savvy computer experts. Because of their diligence and insight, oil is found in places that just a few years ago were regarded as inaccessible. Seismic processing involves the use of complex, innovative products. Most geophysical data processing methods are mathematical and physical hybrids born in the computer and intended to lift explorationists out of the depths of disadvantageous geological conditions. In short, each method, its value derived from the underlying asset of undiscovered oil, is represented by a geophysical model. And geophysical models are only as good as their makers. The issue of relying on computer models to do oil exploration is something we need to look at seriously at all times. The geophysicists make models with whatever is available to them. Are they always able to acquire the correct data that enhance the possibility of finding oil? Do explorationists invariably use the best possible computer models? The geophysicist's fascination with computer models began in the 1950's, when the convolutional model was introduced. Such models introduced the utilization of computer-aided research to uncover oil that was difficult to find. At that time such oil included all the marine prospects.

We must emphasize how important model-testing is. The kinds of geophysical problems we are faced with are in very subtle areas. Geophysicists must look for these kinds of unusual events that require special attention. Long-term reputation for various models can be considerably enhanced by the presence of people who are always looking beyond the current implementations. But some geophysics who have belatedly examined their models might say they are outmoded, when compared with others now in vogue in worldwide processing centers. Moreover, the traditional convolutional model certainly does not take into account the unprecedented recent movements in the evolving geophysical picture. With specification errors mounting, geophysicists apparently hope to recover by speculating on other plays in the exploration program. Of course, what some call speculating, the rest of the world might call gambling, and most gamblers end up ruined. Still, conventional models are able to borrow mightily from past successes and they continue to be used. However some geophysicists say that the collapse of any model should focus attention on the shortcomings of all computer modeling. It is the quality of the unregulated use of conventional products that lies at the heart of the brave new world of modern geophysics. Can it be that geophysicists are not paying attention to the amount of leverage that the use of inappropriate models allow, and should they be more prudent. In the end, any model's brush with ruin can provoked much hand-wringing because the whole geophysical system is built on a framework of mathematical models. As such, the system is fated to tremble from the strains of ever faster and more immense exploration demands directed by the inexact intuitions of explorationists— just the kinds of random stresses that the models cannot predict. When, instead of just trembling, the geophysical system threatens to come unglued, model or no model, mere worry turn to migraines. And fast action is demanded to update the methods. Then the models can settle down again, ripe for a geologic blind-siding by the next bolt from the blue.

Least-squares model

Before an actual digital filter can be designed, it is necessary to set up a model. There are four signals altogether. See Fig. 1. Two signals are required to be known at the outset:

1. The input signal (of length $T + 1$) denoted by

$$x = (x_0, x_1, \cdots, x_T)$$

2. The desired output signal (of length $T + N + 1$) denoted by

GAUSS AND DIGITAL SIGNAL PROCESSING

$$z = (z_0, z_1, \cdots, z_{T+N})$$

The remaining two signals are obtained by means of the least-squares method:

3. The filter (of length $N + 1$) denoted by

$$k = (k_0, k_1, \cdots, k_N)$$

4. The actual output signal (of length $T + N + 1$) denoted by

$$y = (y_0, y_1, \cdots, y_{T+N})$$

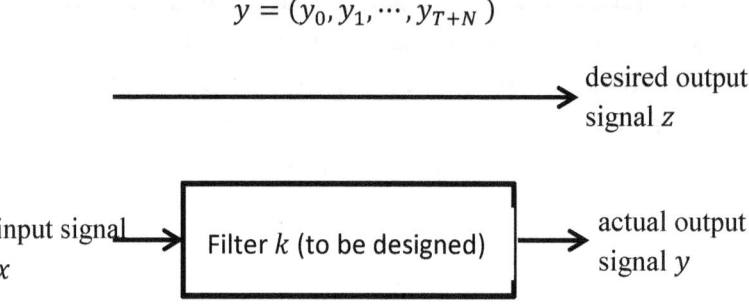

Fig. 1. The least-squares filter design model

Let us now look at some of the properties of this model. The output y is the convolution of the filter with the input; that is,

$$y = k * x$$

More specifically, the output value y_n at time n is given by the discrete convolution formula

$$y_n = \sum_{s=0}^{N} k_s x_{n-s} \quad \text{for } n = 0, 1, \cdots, T + N$$

The input signal and the desired output signal are known. The problem is to find the filter. The Gauss least-squares criterion finds the filter that is optimum in the sense that the sum of squared errors is a minimum.

At time instant n the value of the output is y_n and the value of the desired output is z_n. The difference between the desired output and the actual output is the error $e_n = z_n - y_n$. The *sum of squared errors* is

$$E = \sum_{n=0}^{T+N} (z_n - y_n)^2$$

The *error energy* is another name for the sum of squared errors. We must now find the filter k that minimizes the error energy. The minimum is found by setting the partial derivative of E with respect to each of the filter coefficients equal to zero. The result is the set of a system of N linear simultaneous equations called the normal equations. The *normal equations* are

$$k_0 r_0 + k_1 r_1 + \cdots + k_N r_N = g_0$$
$$k_0 r_1 + k_1 r_0 + \cdots + k_N r_{N-1} = g_1$$
$$\cdots$$
$$k_0 r_N + k_1 r_{N-1} + \cdots + k_N r_0 = g_N$$

The known quantities in the normal equations are

1. the autocorrelation coefficients r_n of the input signal x and
2. the crosscorrelation g_n of the desired output signal z with the input signal x.

The unknown quantities in the normal equations are the values of the coefficients of the filter k.

The autocorrelation of the input is given by

$$r_n = \sum_{i=0}^{T} x_{i+n}\, x_i \quad \text{for } n = 0, 1, \cdots, N$$

and the crosscorrelation between the desired output and the input is given by

$$g_n = \sum_{i=0}^{T} z_{i+n}\, x_i \quad \text{for } n = 0, 1, \cdots, N$$

The solution of the normal equations yields the so-called *least-squares filter k*. The square matrix appearing in the normal equations has the autocorrelation coefficients arranged in Toeplitz form. That is, all the autocorrelation coefficients along any diagonal (the main diagonal or any sub-diagonal) are the same. Because of this Toeplitz structure, the efficient Levinson algorithm can be used in the solution of the normal equations.

Matrix formulation

We now give a matrix formulation for the least-squares filter. This derivation is formulated in terms of matrix theory.

The given signals are:

1. The input signal denoted by the $1 \times T$ column vector

$$x = \begin{pmatrix} x_0 \\ x_1 \\ \ldots \\ x_T \end{pmatrix}$$

2. The desired output signal denoted by the $1 \times (T + N)$ column vector

$$z = \begin{pmatrix} z_0 \\ z_1 \\ \ldots \\ z_{T+N} \end{pmatrix}$$

The computed signals are:

1. The filter denoted by the $(N + 1) \times 1$ column vector

$$k = \begin{pmatrix} k_0 \\ k_1 \\ \ldots \\ k_N \end{pmatrix}$$

2. The actual output signal denoted by the $(T + N + 1) \times 1$ column vector

$$y = \begin{pmatrix} y_0 \\ y_1 \\ \ldots \\ y_{T+N} \end{pmatrix}$$

The quantities X and z are known, and the quantities k and y are unknown.

We now want to express everything in terms of matrices. To ease our task, we will do the matrix formulation for special case when $T = 5$ and $N = 3$. The matrix results that we obtain also apply to matrices associated with any values of T and N. This property is a strong point of matrix theory. The use of matrices avoids much of the bookkeeping.

Because $T + 1 = 6$, the input signal is the 1×6 column vector

Chapter 6. Predictive deconvolution

$$x = \begin{pmatrix} x_0 \\ x_1 \\ x_2 \\ x_3 \\ x_4 \\ x_5 \end{pmatrix}$$

Because $N + T + 1 = 9$, the desired output signal is the 1×9 column vector

$$z = \begin{pmatrix} z_0 \\ z_1 \\ z_2 \\ z_3 \\ z_4 \\ z_5 \\ z_6 \\ z_7 \\ z_8 \end{pmatrix}$$

Because $N + 1 = 4$, the filter is the 4×1 column vector

$$k = \begin{pmatrix} k_0 \\ k_1 \\ k_1 \\ k_3 \end{pmatrix}$$

Because $T + N + 1 = 9$, the actual output signal is the 9×1 column vector

$$y = \begin{pmatrix} y_0 \\ y_1 \\ y_2 \\ y_3 \\ y_4 \\ y_5 \\ y_6 \\ y_7 \\ y_8 \end{pmatrix}$$

Because $(T + N + 1) = 9$ and $(N + 1) \times 4$, we define the 9×4 matrix

GAUSS AND DIGITAL SIGNAL PROCESSING

$$X = \begin{pmatrix} x_0 & 0 & 0 & 0 \\ x_1 & x_0 & 0 & 0 \\ x_2 & x_1 & x_0 & 0 \\ x_3 & x_2 & x_1 & x_0 \\ x_4 & x_3 & x_2 & x_1 \\ x_5 & x_4 & x_3 & x_2 \\ 0 & x_5 & x_4 & x_3 \\ 0 & 0 & x_5 & x_4 \\ 0 & 0 & 0 & x_5 \end{pmatrix}$$

The convolution formula

$$y_n = k_0 x_n + k_1 x_{n-1} + \cdots + k_N x_{n-N} \quad \text{for } n = 0, 1, \cdots, T+N$$

becomes

$$y_n = k_0 x_n + k_1 x_{n-1} + k_2 x_{n-2} + k_3 x_{n-3} \quad \text{for } n = 0, 1, \cdots, 8$$

More explicitly, the convolution is given by the set of individual equations

$$y_0 = k_0 x_0$$
$$y_1 = k_0 x_1 + k_1 x_0$$
$$y_2 = k_0 x_2 + k_1 x_1 + k_2 x_0$$
$$y_3 = k_0 x_3 + k_1 x_2 + k_2 x_1 + k_3 x_0$$
$$y_4 = k_0 x_4 + k_1 x_3 + k_2 x_2 + k_3 x_1$$
$$y_5 = k_0 x_5 + k_1 x_4 + k_2 x_3 + k_3 x_2$$
$$y_6 = k_1 x_5 + k_2 x_4 + k_3 x_3$$
$$y_7 = k_2 x_5 + k_3 x_4$$
$$y_8 = k_3 x_5$$

This set can be expressed as the matrix equation

$$\begin{pmatrix} y_0 \\ y_1 \\ y_2 \\ y_3 \\ y_4 \\ y_5 \\ y_6 \\ y_7 \\ y_8 \end{pmatrix} = \begin{pmatrix} x_0 & 0 & 0 & 0 \\ x_1 & x_0 & 0 & 0 \\ x_2 & x_1 & x_0 & 0 \\ x_3 & x_2 & x_1 & x_0 \\ x_4 & x_3 & x_2 & x_1 \\ x_5 & x_4 & x_3 & x_2 \\ 0 & x_5 & x_4 & x_3 \\ 0 & 0 & x_5 & x_4 \\ 0 & 0 & 0 & x_5 \end{pmatrix} \begin{pmatrix} k_0 \\ k_1 \\ k_1 \\ k_3 \end{pmatrix}$$

Chapter 6. Predictive deconvolution

As a result, the convolution $y = k * x$ can be written as the matrix multiplication

$$y = Xk$$

If we go through the least squares procedure as devised by Gauss, we obtain a set of equations called the *normal equations*. The normal equation involve autocorrelation coefficients of the input. These autocorrelation coefficients are given by

$$r_n = x_n x_0 + x_{1+n} x_1 + \cdots + x_{T+n} x_T \quad \text{for } n = 0, 1, \cdots, N$$

Because $N = 3$ and $T = 5$, the autocorrelation coefficients are given by

$$r_n = x_n x_0 + x_{1+n} x_1 + x_{2+n} x_2 + x_{3+n} x_3 + x_{4+n} x_4 + x_{5+n} x_5$$

$$\text{for } n = 0, 1, 2, 3$$

The set of individual equations is

$$r_0 = x_0 x_0 + x_1 x_1 + x_2 x_2 + x_3 x_3 + x_4 x_4 + x_5 x_5$$

$$r_1 = x_1 x_0 + x_2 x_1 + x_3 x_2 + x_4 x_3 + x_5 x_4$$

$$r_2 = x_2 x_0 + x_3 x_1 + x_4 x_2 + x_5 x_3$$

$$r_3 = x_3 x_0 + x_4 x_1 + x_5 x_2$$

Let the superscript T denote transpose. Because $N + 1 = 4$, define R as the 4×4 autocorrelation matrix

$$R = \begin{pmatrix} r_0 & r_1 & r_2 & r_3 \\ r_1 & r_0 & r_1 & r_2 \\ r_2 & r_1 & r_0 & r_1 \\ r_3 & r_2 & r_1 & r_0 \end{pmatrix}$$

The product $X^T X$ is

$$\begin{pmatrix} x_0 & x_1 & x_2 & x_3 & x_4 & x_5 & 0 & 0 & 0 \\ 0 & x_0 & x_1 & x_2 & x_3 & x_4 & x_5 & 0 & 0 \\ 0 & 0 & x_0 & x_1 & x_2 & x_3 & x_4 & x_5 & 0 \\ 0 & 0 & 0 & x_0 & x_1 & x_2 & x_3 & x_4 & x_5 \end{pmatrix} \begin{pmatrix} x_0 & 0 & 0 & 0 \\ x_1 & x_0 & 0 & 0 \\ x_2 & x_1 & x_0 & 0 \\ x_3 & x_2 & x_1 & x_0 \\ x_4 & x_3 & x_2 & x_1 \\ x_5 & x_4 & x_3 & x_2 \\ 0 & x_5 & x_4 & x_3 \\ 0 & 0 & x_5 & x_4 \\ 0 & 0 & 0 & x_5 \end{pmatrix}$$

If we carry out the matrix multiplication, we see that

GAUSS AND DIGITAL SIGNAL PROCESSING

$$R = X^T X$$

The crosscorrelation between the desired output and the input is given by

$$g_n = \sum_{i=0}^{T} z_{i+n} \, x_i \quad \text{for } n = 0, 1, \cdots, N$$

Because $N = 3$ and $T = 5$, the crosscorrelation coefficients are given by

$$r_n = z_n \, x_0 + z_{1+n} \, x_1 + z_{2+n} \, x_2 + z_{3+n} \, x_3 + z_{4+n} \, x_4 + z_{5+n} \, x_5$$

$$\text{for } n = 0, 1, 2, 3$$

The set of individual equations is

$$g_0 = z_0 \, x_0 + z_1 \, x_1 + z_2 \, x_2 + z_3 \, x_3 + z_4 \, x_4 + z_5 \, x_5$$
$$g_1 = z_1 \, x_0 + z_2 \, x_1 + z_3 \, x_2 + z_4 \, x_3 + z_5 \, x_4 + z_6 \, x_5$$
$$g_2 = z_2 \, x_0 + z_3 \, x_1 + z_4 \, x_2 + z_5 \, x_3 + z_6 \, x_4 + z_7 \, x_5$$
$$g_3 = z_3 \, x_0 + z_4 \, x_1 + z_5 \, x_2 + z_6 \, x_3 + z_7 \, x_4 + z_8 \, x_5$$

The product $X^T z$ is

$$X^T z = \begin{pmatrix} x_0 & x_1 & x_2 & x_3 & x_4 & x_5 & 0 & 0 & 0 \\ 0 & x_0 & x_1 & x_2 & x_3 & x_4 & x_5 & 0 & 0 \\ 0 & 0 & x_0 & x_1 & x_2 & x_3 & x_4 & x_5 & 0 \\ 0 & 0 & 0 & x_0 & x_1 & x_2 & x_3 & x_4 & x_5 \end{pmatrix} \begin{pmatrix} z_0 \\ z_1 \\ z_2 \\ z_3 \\ z_4 \\ z_5 \\ z_6 \\ z_7 \\ z_8 \end{pmatrix}$$

which is

$$\begin{pmatrix} x_0 \, z_0 + x_1 \, z_1 + x_2 \, z_2 + x_3 \, z_3 + x_4 \, z_4 + x_5 \, z_5 \\ x_0 \, z_1 + x_1 \, z_2 + x_2 \, z_3 + x_3 \, z_4 + x_4 \, z_5 + x_5 \, z_6 \\ x_0 \, z_2 + x_1 \, z_3 + x_2 \, z_4 + x_3 \, z_5 + x_4 \, z_6 + x_5 \, z_7 \\ x_0 \, z_3 + x_1 \, z_4 + x_2 \, z_5 + x_3 \, z_6 + x_4 \, z_7 + x_5 \, z_8 \end{pmatrix}$$

We immediately recognize this matrix as g. **Thus we have shown that**

$$g = X^T z$$

The normal equations are

$$R k = g$$

Their solution gives the least-squares filter

Chapter 6. Predictive deconvolution

$$k = R^{-1}g$$

We have thereby determined the filter k. We next must determine the actual output y. We do so by computing

$$y = Xk$$

The signal y is the least squares approximation to the desired output z.

Letting T denote transpose, we can finally compute the sum-squared error by means of the formula

$$E = (z - y)^T(z - y)$$

Numerical example (1) of least-squares filter

Suppose we let

$$x = \begin{pmatrix} 2 \\ -1 \end{pmatrix} \text{ and } z = \begin{pmatrix} 4 \\ -3 \\ 1 \end{pmatrix}$$

Thus $T = 1$ and $T + N = 2$, so $N = 1$. Define the $(T + N + 1) \times (N + 1)$ matrix (i.e., 3 × 2 matrix)

$$X = \begin{pmatrix} 2 & 0 \\ -1 & 2 \\ 0 & -1 \end{pmatrix}$$

It follows that

$$R = X^T X = \begin{pmatrix} 2 & -1 & 0 \\ 0 & 2 & -1 \end{pmatrix} \begin{pmatrix} 2 & 0 \\ -1 & 2 \\ 0 & -1 \end{pmatrix} = \begin{pmatrix} 5 & -2 \\ -2 & 5 \end{pmatrix}$$

Compute g by the equation

$$g = X^T z = \begin{pmatrix} 2 & -1 & 0 \\ 0 & 2 & -1 \end{pmatrix} \begin{pmatrix} 4 \\ -3 \\ 1 \end{pmatrix} = \begin{pmatrix} 11 \\ -7 \end{pmatrix}$$

The solution of the normal equations gives the filter as

$$k = R^{-1}g = \begin{pmatrix} 5 & 2 \\ 2 & 5 \end{pmatrix}^{-1} \begin{pmatrix} 11 \\ -7 \end{pmatrix} = \begin{pmatrix} \frac{23}{7} \\ -\frac{19}{7} \end{pmatrix}$$

The least squares approximation y is

$$y = Xk = \begin{pmatrix} 2 & 0 \\ -1 & 2 \\ 0 & -1 \end{pmatrix} \begin{pmatrix} \frac{23}{7} \\ \frac{19}{7} \end{pmatrix} = \begin{pmatrix} \frac{46}{7} \\ -\frac{61}{7} \\ \frac{19}{7} \end{pmatrix}$$

We have

$$z - y = \begin{pmatrix} 4 \\ -3 \\ 1 \end{pmatrix} - \begin{pmatrix} \frac{46}{7} \\ -\frac{61}{7} \\ \frac{19}{7} \end{pmatrix} = \begin{pmatrix} -\frac{18}{7} \\ \frac{40}{7} \\ -\frac{12}{7} \end{pmatrix}$$

The error energy is

$$E = (z - y)^T(z - y) = \begin{pmatrix} -\frac{18}{7} & \frac{40}{7} & -\frac{12}{7} \end{pmatrix} \begin{pmatrix} -\frac{18}{7} \\ \frac{40}{7} \\ -\frac{12}{7} \end{pmatrix} = \left(\frac{2068}{49}\right) = 42.2$$

Numerical example (2) of least-squares filter

Suppose we let

$$x = \begin{pmatrix} 2 \\ 1 \end{pmatrix} \text{ and } z = \begin{pmatrix} 1 \\ -0.5 \\ 0 \end{pmatrix}$$

Thus $T = 1$ and $T + N = 2$, so $N = 1$. Define the 3 × 2 matrix

$$X = \begin{pmatrix} 2 & 0 \\ 1 & 2 \\ 0 & 1 \end{pmatrix}$$

It follows that

$$R = X^T X = \begin{pmatrix} 2 & 1 & 0 \\ 0 & 2 & 1 \end{pmatrix} \begin{pmatrix} 2 & 0 \\ 1 & 2 \\ 0 & 1 \end{pmatrix} = \begin{pmatrix} 5 & 2 \\ 2 & 5 \end{pmatrix}$$

Define g as

$$g = X^T z = \begin{pmatrix} 2 & 1 & 0 \\ 0 & 2 & 1 \end{pmatrix} \begin{pmatrix} 1 \\ -0.5 \\ 0 \end{pmatrix} = \begin{pmatrix} 1.5 \\ -1 \end{pmatrix}$$

Chapter 6. Predictive deconvolution

The solution of the normal equations gives

$$k = R^{-1}g = \begin{pmatrix} 5 & 2 \\ 2 & 5 \end{pmatrix}^{-1} \begin{pmatrix} 1.5 \\ -1 \end{pmatrix} = \begin{pmatrix} \frac{19}{42} \\ -\frac{8}{21} \end{pmatrix}$$

The least squares approximation y is

$$y = Xk = \begin{pmatrix} 2 & 0 \\ 1 & 2 \\ 0 & 1 \end{pmatrix} \begin{pmatrix} \frac{19}{42} \\ -\frac{8}{21} \end{pmatrix} = \begin{pmatrix} \frac{19}{21} \\ -\frac{13}{42} \\ -\frac{8}{21} \end{pmatrix}$$

We have

$$z - y = \begin{pmatrix} 1 \\ -0.5 \\ 0 \end{pmatrix} - \begin{pmatrix} \frac{19}{21} \\ -\frac{13}{42} \\ -\frac{8}{21} \end{pmatrix} = \begin{pmatrix} \frac{2}{21} \\ -\frac{4}{21} \\ \frac{8}{21} \end{pmatrix}$$

The error energy is

$$E = (z - y)^T (z - y) = \left(\frac{4}{21}\right) = 0.190$$

Prediction filter

Let us now examine the filter that predicts the future values of a signal from past values of the signal. We denote the prediction filter by the $1 \times N$ row vector

$$k = (k_0, k_1, k_2, \cdots, k_{N-1})$$

The input to the filter is the signal x_n and the desired output is the time-advanced version of the input. The prediction distance is given by the positive integer α. The input to the filter is the signal x_n at present time n and the desired output z_n is the signal $x_{n+\alpha}$ at future time $n + \alpha$. The filter is designed so that the output y_n at the present time n is an least-squares estimate of the future value $x_{n+\alpha}$. Let this estimate is denoted by $\hat{x}_{n+\alpha}$ so as to distinguish it from the actual value $x_{n+\alpha}$. The action of the filter is represented by the convolution

$$\hat{x}_{n+a} = k_0 x_n + k_1 x_{n-1} + k_2 x_{n-2} + \cdots + k_{N-1} x_{n-N+1}$$

This equation says that the prediction operator acts on an input up to time n and estimates the its value at the future time $n + \alpha$.

The normal equations can be used to compute the filter coefficients. The right hand side of the normal equations is given by the crosscorrelation between the time-advanced trace and the trace. More precisely, the time-advanced trace is

$$z_i = x_{i+\alpha}$$

The equation

$$g_n = \sum_{i=0}^{T} z_{i+n} x_i$$

becomes

$$g_n = \sum_{i=0}^{T} x_{i+\alpha+n} x_i = r_{n+\alpha}$$

As the above equation shows, the required crosscorrelation between the desired output and the input is equal to the autocorrelation of the input for lags greater than or equal to α. Thus the normal equations for the prediction filter may be written

$$\begin{pmatrix} r_0 & r_1 & \cdots & r_{N-1} \\ r_1 & r_0 & \cdots & r_{N-2} \\ \cdots & \cdots & \cdots & \cdots \\ r_{N-1} & r_{N-2} & \cdots & r_0 \end{pmatrix} \begin{pmatrix} k_0 \\ k_1 \\ \cdots \\ k_{N-1} \end{pmatrix} = \begin{pmatrix} r_\alpha \\ r_{\alpha+1} \\ \cdots \\ r_{\alpha+N-1} \end{pmatrix}$$

Their solution yields the coefficients of the optimum prediction operator for prediction distance α.

Prediction-error filter for unit prediction distance

We will let p designate the prediction-error filter for unit prediction distance. Whereas prediction filters can be used with various values of the prediction distance, attention now turns to the prediction filter for unit prediction distance, that is, for $\alpha = 1$. The prediction filter for unit prediction distance is

$$\hat{x}_n = f_0 x_{n-1} + f_1 x_{n-1} + \cdots + f_{N-1} x_{n-N+1}$$

Chapter 6. Predictive deconvolution

Setting $\alpha = 1$ in the normal equations in the previous section, we obtain the normal equations

$$\begin{pmatrix} r_0 & r_1 & \cdots & r_{N-1} \\ r_1 & r_0 & \cdots & r_{N-2} \\ \cdots & \cdots & \cdots & \cdots \\ r_{N-1} & r_{N-2} & \cdots & r_0 \end{pmatrix} \begin{pmatrix} f_0 \\ f_1 \\ \cdots \\ f_{N-1} \end{pmatrix} = \begin{pmatrix} r_1 \\ r_2 \\ \cdots \\ r_N \end{pmatrix}$$

The solution of these normal equations gives the coefficients of the prediction filter $k = (f_0, f_1, \cdots, f_{N-1})$ for unit prediction distance.

We have just defined the prediction filter for unit prediction distance. Now let us define the corresponding prediction error filter. We rewrite the above equation as

$$\begin{pmatrix} r_1 \\ r_2 \\ \cdots \\ r_N \end{pmatrix} + \begin{pmatrix} r_0 & r_1 & \cdots & r_{N-1} \\ r_1 & r_0 & \cdots & r_{N-2} \\ \cdots & \cdots & \cdots & \cdots \\ r_{N-1} & r_{N-2} & \cdots & r_0 \end{pmatrix} \begin{pmatrix} -f_0 \\ -f_1 \\ \cdots \\ -f_{N-1} \end{pmatrix} = \begin{pmatrix} 0 \\ 0 \\ \cdots \\ 0 \end{pmatrix}$$

which is

$$\begin{pmatrix} r_1 & r_0 & r_1 & \cdots & r_{N-1} \\ r_2 & r_1 & r_0 & \cdots & r_{N-2} \\ \cdots & \cdots & \cdots & \cdots & \cdots \\ r_N & r_{N-1} & r_{N-2} & \cdots & r_0 \end{pmatrix} \begin{pmatrix} 1 \\ -f_0 \\ -f_1 \\ \cdots \\ -f_{N-1} \end{pmatrix} = \begin{pmatrix} 0 \\ 0 \\ \cdots \\ 0 \end{pmatrix}$$

Next define ρ by the equation

$$(r_0 \quad r_1 \quad \cdots \quad r_N) \begin{pmatrix} 1 \\ -f_0 \\ \cdots \\ -f_{N-1} \end{pmatrix} = r_0 - f_0 r_1 - \cdots - f_{N-1} r_N = \rho$$

We combine the above two equation to obtain

$$\begin{pmatrix} r_0 & r_1 & \cdots & r_N \\ r_1 & r_0 & \cdots & r_{N-1} \\ \cdots & \cdots & \cdots & \cdots \\ r_N & r_{N-1} & \cdots & r_0 \end{pmatrix} \begin{pmatrix} 1 \\ -f_0 \\ \cdots \\ -f_{N-1} \end{pmatrix} = \begin{pmatrix} \rho \\ 0 \\ \cdots \\ 0 \end{pmatrix}$$

The column vector on the left hand side is the prediction operator for unit prediction distance; that is

$$p = \begin{pmatrix} 1 \\ -f_0 \\ \ldots \\ -f_{N-1} \end{pmatrix}$$

The prediction error filter for unit prediction distance is necessarily minimum-phase (Robinson and Wold, 1963).

Numerical example of prediction error filter

Let $T = 3$ and $N = 1$. Because $T + 1 = 3$, the input signal is a 3×1 vector, which we take to be

$$x = \begin{pmatrix} 3 \\ -2 \\ 1 \\ 0 \end{pmatrix}$$

The desired output is the input signal advanced by one time unit. Because $N + T + 1 = 5$, the desired output is the 5×1 vector

$$z = \begin{pmatrix} -2 \\ 1 \\ 0 \\ 0 \\ 0 \end{pmatrix}$$

Because $N = 1$, the required prediction filter is the 2×1 vector

$$f = \begin{pmatrix} f_0 \\ f_1 \end{pmatrix}$$

Because $(M + N + 1) = 5$ and $(M + 1) = 2$, we have the 5×2 input matrix

$$X = \begin{pmatrix} 3 & 0 \\ -2 & 3 \\ 1 & -2 \\ 0 & 1 \\ 0 & 0 \end{pmatrix}$$

The autocorrelation matrix is

$$X^T X = R$$

which gives

Chapter 6. Predictive deconvolution

$$R = \begin{pmatrix} 3 & -2 & 1 & 0 & 0 \\ 0 & 3 & -2 & 1 & 0 \end{pmatrix} \begin{pmatrix} 3 & 0 \\ -2 & 3 \\ 1 & -2 \\ 0 & 1 \\ 0 & 0 \end{pmatrix} = \begin{pmatrix} 14 & -8 \\ -8 & 14 \end{pmatrix}$$

The right hand side of the normal equations is

$$X^T z = g$$

which is

$$g = \begin{pmatrix} 3 & -2 & 1 & 0 & 0 \\ 0 & 3 & -2 & 1 & 0 \end{pmatrix} \begin{pmatrix} -2 \\ 1 \\ 0 \\ 0 \\ 0 \end{pmatrix} = \begin{pmatrix} -8 \\ 3 \end{pmatrix}$$

Then the normal equations may be written as

$$Rf = g$$

The required filter is

$$R^{-1}g = f$$

which is

$$f = \begin{pmatrix} 14 & -8 \\ -8 & 14 \end{pmatrix}^{-1} \begin{pmatrix} -8 \\ 3 \end{pmatrix} = \begin{pmatrix} -\frac{2}{3} \\ \frac{1}{6} \\ -\frac{1}{6} \end{pmatrix}$$

The prediction error filter for unit prediction distance is

$$p = \begin{pmatrix} 1 \\ -f_0 \\ -f_1 \end{pmatrix} = \begin{pmatrix} 1 \\ \frac{2}{3} \\ \frac{1}{6} \end{pmatrix}$$

The value of p is

$$\rho = (r_0 \quad r_1 \quad r_2) \begin{pmatrix} 1 \\ -f_0 \\ -f_1 \end{pmatrix}$$

From

$$R = \begin{pmatrix} 14 & -8 \\ -8 & 14 \end{pmatrix}$$

we know that $r_0 = 14$ and $r_1 = -8$. However, we still need r_2. We compute

$$r_2 = \begin{pmatrix} 3 & -2 & 1 & 0 & 0 & 0 \end{pmatrix} \begin{pmatrix} 0 \\ 0 \\ 3 \\ -2 \\ 1 \\ 0 \end{pmatrix} = 3$$

Thus the value of ρ is

$$\rho = \begin{pmatrix} r_0 & r_1 & r_2 \end{pmatrix} \begin{pmatrix} 1 \\ -f_0 \\ -f_1 \end{pmatrix} = \begin{pmatrix} 14 & -8 & 3 \end{pmatrix} \begin{pmatrix} 1 \\ \frac{2}{3} \\ \frac{1}{6} \end{pmatrix} = \frac{55}{6}$$

As we have seen in the previous section, the normal equations for the prediction error filter for unit prediction distance are

$$\begin{pmatrix} r_0 & r_1 & \cdots & r_N \\ r_1 & r_0 & \cdots & r_{N-1} \\ \vdots & \vdots & \cdots & \vdots \\ r_N & r_{N-1} & \cdots & r_0 \end{pmatrix} \begin{pmatrix} 1 \\ -f_0 \\ \vdots \\ -f_{N-1} \end{pmatrix} = \begin{pmatrix} \rho \\ 0 \\ \vdots \\ 0 \end{pmatrix}$$

In this case, the normal equations are

$$\begin{pmatrix} 14 & -8 & 3 \\ -8 & 14 & -8 \\ 3 & -8 & 14 \end{pmatrix} \begin{pmatrix} p_0 \\ p_1 \\ p_2 \end{pmatrix} = \begin{pmatrix} 55 \\ 6 \\ 0 \\ 0 \end{pmatrix}$$

with solution

$$\begin{pmatrix} p_0 \\ p_1 \\ p_2 \end{pmatrix} = \begin{pmatrix} 1 \\ -f_0 \\ \vdots \\ -f_{N-1} \end{pmatrix} = \begin{pmatrix} 1 \\ \frac{2}{3} \\ \frac{1}{6} \end{pmatrix}$$

Spiking filter

Let us now find the operator that shapes the input into a unit spike. In other words, the desired output is

$$z = (1, 0, \cdots, 0)$$

The right-hand side of the normal equations is given by

Chapter 6. Predictive deconvolution

$$g_n = \sum_{i=0}^{T} z_{i+n} x_i \quad \text{for } n = 0, 1, \cdots, N$$

which is

$$g_0 = z_0 x_0 + z_1 x_1 + \cdots + z_T x_T$$
$$g_1 = z_1 x_0 + z_2 x_1 + \cdots + z_T x_{T-1}$$
$$\cdots$$
$$g_N = z_N x_0 + z_{N+1} x_1 + \cdots + z_T x_{T-N}$$

However, all the values of z_i are zero except $z_0 = 1$. Thus

$$g_0 = x_0$$
$$g_1 = 0$$
$$\cdots$$
$$g_N = 0$$

Thus the normal equations are

$$\begin{pmatrix} r_0 & r_1 & \cdots & r_N \\ r_1 & r_0 & \cdots & r_{N-1} \\ \cdots & \cdots & \cdots & \cdots \\ r_N & r_{N-1} & \cdots & r_0 \end{pmatrix} \begin{pmatrix} a_0 \\ a_1 \\ \cdots \\ a_N \end{pmatrix} = \begin{pmatrix} x_0 \\ 0 \\ \cdots \\ 0 \end{pmatrix}$$

Numerical example of spiking filter

Let $T = 3$ and $N = 2$. Because $T + 1 = 3$, the input signal is a 3×1 vector, which we take to be

$$x = \begin{pmatrix} 3 \\ -2 \\ 1 \\ 0 \end{pmatrix}$$

The desired output is the input signal advanced by one time unit. Because $N + T + 1 = 6$, the desired output is the 6×1 spike

$$z = \begin{pmatrix} 1 \\ 0 \\ 0 \\ 0 \\ 0 \\ 0 \end{pmatrix}$$

Let the required spiking filter (with $M = 2$) be

GAUSS AND DIGITAL SIGNAL PROCESSING

$$a = \begin{pmatrix} a_0 \\ a_1 \\ a_2 \end{pmatrix}$$

Define the $(T + N + 1) \times (N + 1)$ matrix or 6×3 matrix

$$X = \begin{pmatrix} 3 & 0 & 0 \\ -2 & 3 & 0 \\ 1 & -2 & 3 \\ 0 & 1 & -2 \\ 0 & 0 & 1 \\ 0 & 0 & 0 \end{pmatrix}$$

The $(M + 1) \times (M + 1)$ matrix R is

$$X^T X = R$$

which is

$$\begin{pmatrix} 3 & -2 & 1 & 0 & 0 & 0 \\ 0 & 3 & -2 & 1 & 0 & 0 \\ 0 & 0 & 3 & -2 & 1 & 0 \end{pmatrix} \begin{pmatrix} 3 & 0 & 0 \\ -2 & 3 & 0 \\ 1 & -2 & 3 \\ 0 & 1 & -2 \\ 0 & 0 & 1 \\ 0 & 0 & 0 \end{pmatrix} = \begin{pmatrix} 14 & -8 & 3 \\ -8 & 14 & -8 \\ 3 & -8 & 14 \end{pmatrix}$$

The right hand side is

$$X^T z = g$$

which is

$$\begin{pmatrix} 3 & -2 & 1 & 0 & 0 & 0 \\ 0 & 3 & -2 & 1 & 0 & 0 \\ 0 & 0 & 3 & -2 & 1 & 0 \end{pmatrix} \begin{pmatrix} 1 \\ 0 \\ 0 \\ 0 \\ 0 \\ 0 \end{pmatrix} = \begin{pmatrix} 3 \\ 0 \\ 0 \end{pmatrix}$$

Then the normal equations may be written as

$$Ra = g$$

which is

$$\begin{pmatrix} 14 & -8 & 3 \\ -8 & 14 & -8 \\ 3 & -8 & 14 \end{pmatrix} \begin{pmatrix} a_0 \\ a_1 \\ a_2 \end{pmatrix} = \begin{pmatrix} 3 \\ 0 \\ 0 \end{pmatrix}$$

For the spiking filter, the solution of the normal equations is

Chapter 6. Predictive deconvolution

$$a = \begin{pmatrix} a_0 \\ a_1 \\ a_2 \end{pmatrix} = \begin{pmatrix} 14 & -8 & 3 \\ -8 & 14 & -8 \\ 3 & -8 & 14 \end{pmatrix}^{-1} \begin{pmatrix} 3 \\ 0 \\ 0 \end{pmatrix} = \begin{pmatrix} \frac{18}{55} \\ \frac{12}{55} \\ \frac{3}{55} \end{pmatrix}$$

Let us now compare the spiking filter to the prediction error filter for unit prediction distance. For the latter, the solution of the normal equations is

$$p = \begin{pmatrix} p_0 \\ p_1 \\ p_2 \end{pmatrix} = \begin{pmatrix} 14 & -8 & 3 \\ -8 & 14 & -8 \\ 3 & -8 & 14 \end{pmatrix}^{-1} \begin{pmatrix} 55 \\ 6 \\ 0 \\ 0 \end{pmatrix} = \begin{pmatrix} 1 \\ \frac{2}{3} \\ \frac{1}{6} \end{pmatrix}$$

Thus the two filters are related by a scale factor. Since $3 = x_0$ and $55/6 = \rho$, the scale factor to go from p to a is

$$\frac{x_0}{\rho} = \frac{3}{\left(\frac{55}{6}\right)} = \frac{18}{55}$$

which gives

$$a_0 = \frac{18}{55} 1 = \frac{18}{55}, \quad a_1 = \frac{18}{55} \frac{2}{3} = \frac{12}{55}, \quad a_2 = \frac{18}{55} \frac{1}{6} = \frac{3}{55}$$

Chapter 7. Seismic waves

Gauss: All the measurements in the world do not balance *one* theorem by which the science of eternal truths is actually advanced.

Dual fields

From Newton's second law, it is possible to directly derive the following first-order partial differential equation

$$\frac{\partial p}{\partial z} = -\rho \frac{\partial V}{\partial t} \qquad (1)$$

The variable z is the vertical axis and t is the time. The function p is the pressure and the function V is the particle velocity. The density, or mass per unit volume, is denoted by ρ. A symmetric first-order partial differential equation can be obtained from Hooke's law

$$\frac{\partial p}{\partial t} = -K \frac{\partial V}{\partial z} \qquad (2)$$

where K is Young's modulus. By cross differentiation of (1) and (2), the pressure and particle velocity each obey the wave equation:

$$\frac{1}{c^2} \frac{\partial^2 p}{\partial t^2} = \frac{\partial^2 p}{\partial z^2} \quad \text{and} \quad \frac{1}{c^2} \frac{\partial^2 V}{\partial t^2} = \frac{\partial^2 V}{\partial z^2}$$

where the propagation velocity c obeys the relation $K = \rho c^2$.

For non-dissipative and non-dispersive electromagnetic waves, Maxwell's equations yield

$$\nabla \times E = -\mu \frac{\partial H}{\partial t} \quad \text{and} \quad \nabla \times E = -\varepsilon \frac{\partial E}{\partial t}$$

where E and H are the electric and magnetic fields and ε and μ are the dielectric and permeability constants, respectively. For horizontal polarization of $E = (E_x, 0, 0)$, and $H = (0, H_y, 0)$, Maxwell's equations give the symmetric first-order partial differential equations

$$\frac{\partial E_x}{\partial z} = -\mu \frac{\partial H_y}{\partial t} \quad \text{and} \quad \frac{\partial E_x}{\partial t} = -\frac{1}{\varepsilon} \frac{\partial H_y}{\partial t}$$

Again by cross differentiation, it is seen that E_x and H_y each satisfy the wave equation

$$\frac{1}{c^2}\frac{\partial^2 E_x}{\partial t^2} = \frac{\partial^2 E_x}{\partial z^2} \quad \text{and} \quad \frac{1}{c^2}\frac{\partial^2 H_y}{\partial t^2} = \frac{\partial^2 H_y}{\partial z^2}$$

where the propagation velocity c of light obeys the relation

$$c = \frac{1}{\sqrt{\varepsilon\mu}}$$

Dual sensor

The seismic method relies on the fact that seismic traveling waves propagate through the earth. In reflection seismic surveying, seismic waves are generated at shot points and the waves travel downward through the earth. As the downward-traveling waves encounter various reflecting surfaces between successive subsurface layers, they are partially reflected upward. The resulting upward-traveling waves are also partially reflected downward when they encounter reflecting surfaces between successive layers. The waves received at buried or submerged detection points are composed of both upward-traveling waves and downward-traveling waves.

A dual sensor consists of both a pressure-sensitive seismic detector and a velocity-sensitive seismic detector. The outputs of the two detectors are combined order to attain separately the upcoming waves and the downgoing traveling waves. In other words, the dual sensor renders the receiving system suitable for differentiating between various waves, such as longitudinal waves traveling in one direction and longitudinal waves traveling in the opposite direction.

Virtually all of geophysical signal processing is based on ray theory. A wavefield represents the spatial perturbations at a given time that result from the passage of a traveling wave, such as the pressure changes caused by a traveling seismic wave. A wavefield may be described as the temporal perturbations that result from traveling seismic waves. For example, seismic migration is sometimes referred to as downward propagation of the wavefield. A waveform is a plot (usually as a function of time) of a quantity involved in wave motion, such as voltage, current, seismic displacement, etc. A wavefront is the surface over which the phase of a traveling wave disturbance is the same. The wavefront moves perpendicular to itself as the wave disturbance travels in an isotropic medium. A wavefront may be described as a locus of equal traveltime. Usually we think of a wavefront as the leading edge of a waveform.

A raypath is a line everywhere perpendicular to wavefronts (in isotropic media). While seismic energy does not travel only along raypaths (i.e., seismic energy would reach points by diffraction even if the raypaths were blocked), raypaths constitute a useful method of determining arrival time by ray tracing. Raypaths are usually shown on wavefront charts.

In order to effectively use all of the seismic processing methods available today, it is essential to be able to measure the traveling waves upon which ray theory is based. A hydrophone cannot measure a traveling wave as such. Neither can a geophone. However, in a dual sensor, both types of detectors can be used in conjunction with one another to measure the traveling waves required for effective geophysical signal processing. The dual sensor involves two detectors. . One seismic wave detector at each receiving station is pressure-sensitive, while the other seismic wave detector located at the same receiving station is velocity-sensitive.

We use the symbol V for particle velocity and the symbol p for pressure. We let D denote the downgoing component of the particle velocity disturbance and let U denote the upgoing component of the particle velocity disturbance. Similarly, we let d denote the downgoing component of the pressure disturbance and let u denote the upgoing component of the pressure disturbance.

Augustin Jean Fresnel (1788-1827) showed that the definition of the reflection coefficient of an interface requires the consideration of both the particle velocity attribute and the pressure attribute of the wave motion on each side of the interface. Both of these attributes must be continuous across the interface. A disturbance satisfying the wave equation is equal to the sum of two traveling waves, one of which travels downward and the other upward. The downgoing and upgoing waves transport the energy to and from a reflecting horizon, and the particle velocity and the pressure attributes of Fresnel determine the partition of energy at that horizon.

The net result is that seismic processing simply cannot be done with conventionally recorded data unless one is willing to make traveling-wave assumptions. However, no traveling-wave assumption has to be made if a dual-sensor (that is, both a geophone and a hydrophone) is used as the receiver. A dual sensor provides both the hydrophone signal and the geophone signal. Let us give an example. Look at Fig. 19. The geophone and hydrophone signals are out-of-phase. As a result it can be concluded that the event is a traveling upgoing wave. On the other hand, if the geophone and hydrophone signals were in-phase, then the event would be a traveling downgoing wave.

Let us consider the case of a buried receiver. The source can also be buried, but it should placed at a level above the level of the receiver. The layered earth system (lying above the basement rock) is then divided into two subsystems by the horizontal plane the passes through the receiver. The upper subsystem of layers is called the shallow system and the lower subsystem is called the deep system. Because the shallow system contains the source of energy, it is active. Because the deep system does not contain a source of energy, it is passive. See Fig. 1.

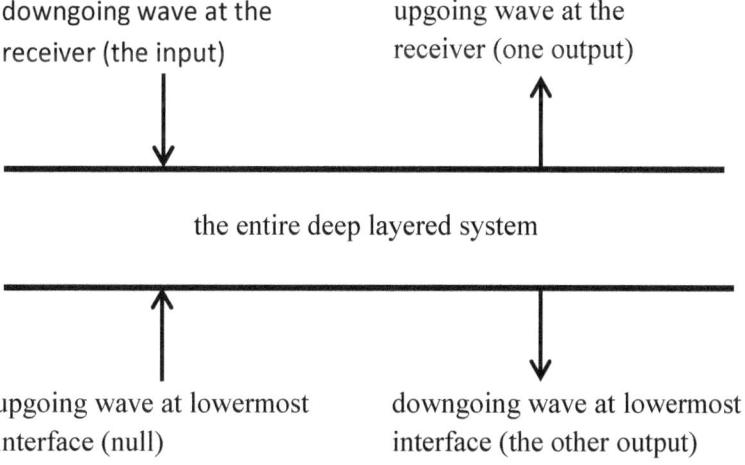

Fig. 1. The deep system has two inputs and two outputs

In the deep system, the downgoing wave at the receiver is one input. The upgoing wave at the lowermost interface is the other input (which is null). The upgoing wave at the receiver is one output. The downgoing wave at the lowermost interface is the other output. The deep system is a passive system with only one input (the downgoing wave at the receiver). If we consider the output given by the upgoing wave at the receiver, then we have a conventional convolutional model with one input and one output. See Fig. 2.

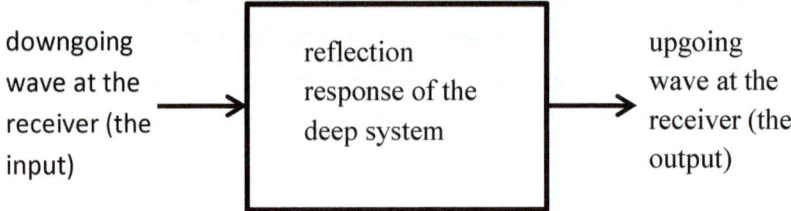

Fig. 2. The convolutional model for the deep-layered system

The reflection response involves just the reflection coefficients of the deep system, and in no way depends upon the reflection coefficients of the shallow system. Two wavelets have to be estimated, namely the input wavelet and the output wavelet. Both occur at the receiver location.

A geophone measures the particle-velocity attribute of the seismic disturbance, and a hydrophone measures the pressure attribute of the seismic disturbance. The seismic processing is based upon the availability of downgoing and upgoing traveling waves. However, a traveling wave is never recorded as such in seismic acquisition. The signal recorded by a geophone is the sum of the particle velocity attributes of the downgoing and upgoing waves. The signal recorded by a hydrophone is the sum of the pressure attributes of the downgoing and upgoing waves.

When the receiver is placed at the surface of the earth, we naturally assumed that the received wave was upgoing. This assumption is called a traveling wave assumption. However no such assumption is possible when the receiver is placed below the surface.

What can we do? We must use a dual sensor and record both the particle velocity attribute and the pressure attribute. Then we must go back in time and appeal to Augustin Jean Fresnel (1788-1827) and Jean Le Rond d'Alembert (1717-1783) for their help.

Augustin Jean Fresnel (1788-1827) showed that the determination of the reflection coefficient of an interface requires the consideration of both the particle velocity attribute and the pressure attribute of the wave motion on each side of the interface. Both of these attributes must be continuous across the interface.

Jean Le Rond d'Alembert (1717-1783) showed that a disturbance satisfying the wave equation is equal to the sum of two traveling waves, one of which travels downward and the other upward. The downgoing and upgoing waves of d'Alembert transport the energy to and from a reflecting horizon, and the particle velocity and the pressure attributes of Fresnel determine the partition of energy at that horizon.

A traveling-wave assumption is not needed if a dual-sensor (that is, both a geophone and a hydrophone) is used as the receiver. A dual sensor provides both the hydrophone signal and the geophone signal. Let us give an example.

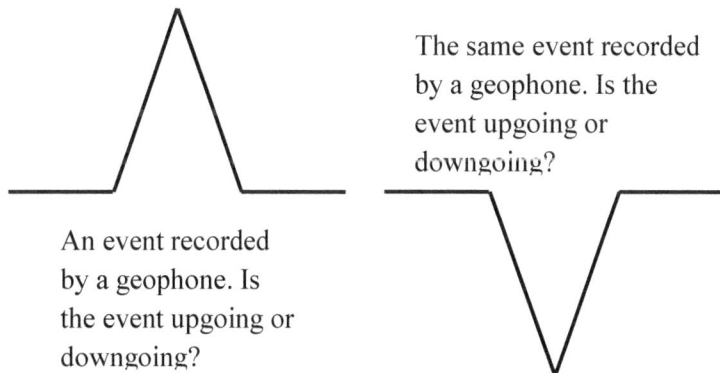

An event recorded by a geophone. Is the event upgoing or downgoing?

The same event recorded by a geophone. Is the event upgoing or downgoing?

Fig. 3. The event is a traveling upgoing wave

Look at Fig. 3. The geophone and hydrophone signals are out-of-phase. Neither the geophone signal by itself, nor the hydrophone signal by itself, is enough to determine whether the event is upgoing or downgoing. However the two signals are out of phase, so (by the convention used here) the event is upgoing. On the other hand, if the geophone and hydrophone signals were in-phase, then the event would be a traveling downgoing wave.

d'Alembert equations

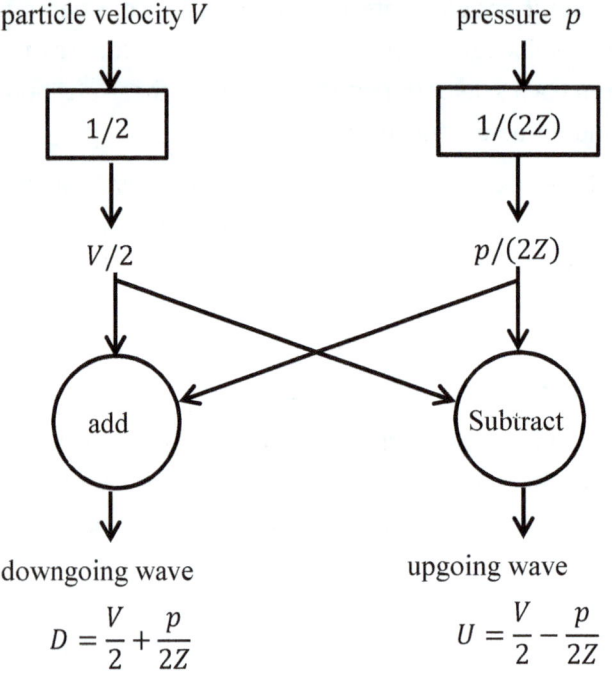

Fig. 4. The d'Alembert equations

In the case of dual sensors, the traveling waves can be computed directly from the data by means of the d'Alembert equations. See. Fig. 4. The inputs are the particle velocity, pressure, and acoustic impedance, all measured at the receiver location. The outputs are the downgoing particle-velocity wave and the upgoing particle-velocity wave, both occurring at the receiver location.

The receiver is a dual sensor buried below the source of seismic energy. The dual sensor measures the particle velocity signal and the pressure signal at the receiver location. By use of the d'Alembert equations, the particle velocity signal and the pressure signal are converted into the downgoing wave and the upgoing wave at the receiver location. The downgoing wave is the input signal and the upgoing wave is the output signal that occurs in the convolutional model (as shown in Fig. 2) of the deep system.

We can derive the d'Alembert equations as follows. For a given rock layer, let ρ denote the density and let v denote the wave propagation velocity. The product $Z = \rho v$ is the acoustic impedance. A dual sensor is made up of a geophone and a hydrophone. The geophone records the particle-velocity trace V and the hydrophone records the pressure trace p. Each trace is equal to the sum of the downgoing wave motion plus the upcoming wave motion at the sensor. Let D denote the downgoing wave motion of the particle-velocity trace and let U denote the upgoing wave motion of the particle velocity trace. Similarly, let d denote the downgoing wave motion of the pressure trace and let u denote the upgoing wave motion of the pressure trace. Thus we have the two equations (the first equation for the particle velocity trace and the second for the pressure trace):

$$V = D + U \quad \text{and} \quad p = d + u$$

There are various conventions used. We will use the Berkhout convention.

(1) The downgoing wave motion d has the same polarity as the downgoing wave motion D and that the two are related by a scale factor given by the acoustic impedance.

(2) The upgoing wave motion u has the opposite polarity as the upgoing wave motion U and that the same scale factor relates the two.

Thus we have

$$d = ZD \quad \text{and} \quad u = -ZU$$

The solution of the above equations yields the d'Alembert equations for the downgoing and upgoing particle-velocity wave motion

$$D = \frac{V + p/Z}{2} \quad \text{and} \quad U = \frac{V - p/Z}{2}$$

We could alternatively solve for the downgoing and upgoing pressure wave motion to obtain the corresponding d'Alembert equations

$$d = \frac{ZV + p}{2} \quad \text{and} \quad u = \frac{-ZV + p}{2}$$

Einstein deconvolution

For any method of deconvolution, the output signal is known. In general, deconvolution consists of two steps. The first step is to determine the input signal. The second step is to deconvolve the output signal by the input signal to yield the unit-impulse response.

In signature deconvolution, the input wavelet (the signature) is measured directly. In spiking deconvolution, the input wavelet is estimated from knowledge of the output signal.

In the case when the receiver is a dual-sensor, the d'Alembert equations yield the input signal and the output signal. Thus we can apply Einstein deconvolution; namely, the deconvolution of the output signal by the input signal. The mathematics is implicit in Einstein's special theory of relativity.

The main difference between Einstein deconvolution and spiking deconvolution is in the first step; namely, in how the input signal is obtained. Spiking deconvolution can be used in cases where the input signal can be neither directly measured nor directly estimated. In such cases the input signal must be determined indirectly.

In order to do Einstein deconvolution, a dual geophone/hydrophone sensor is required to obtain the downgoing input signal and upgoing output signal. The d'Alembert equations are used, so no traveling-wave assumption is required. Einstein deconvolution removes all the effects that are introduced by everything above the receiver location. Einstein deconvolution not only eliminates the source signature, but also eliminates the ghosts and reverberations due to the layers above the receiver. Like conventional signature deconvolution, Einstein deconvolution requires knowledge of the input signal as well as the output signal. The input wavelet is the downgoing wave at the receiver and the output signal is the upgoing wave at the receiver. The deconvolved signal is an estimate of the reflection response of the deep system (that is, the earth layering below the receiver).

The Einstein deconvolution method can be described in two steps. See Fig. 5. The first step is to convert the particle velocity and pressure signals into the downgoing wave and the upgoing wave at the location of the receiver. In order to carry out this step the d'Alembert equations are used. The second step is to deconvolve the upgoing wave by the downgoing wave. Einstein deconvolution removes the (unknown and not necessarily minimum-delay) source signature as well as the reverberations and ghosts due to the layers above the receiver. The result of Einstein deconvolution is the unit-impulse reflection response of deep earth system.

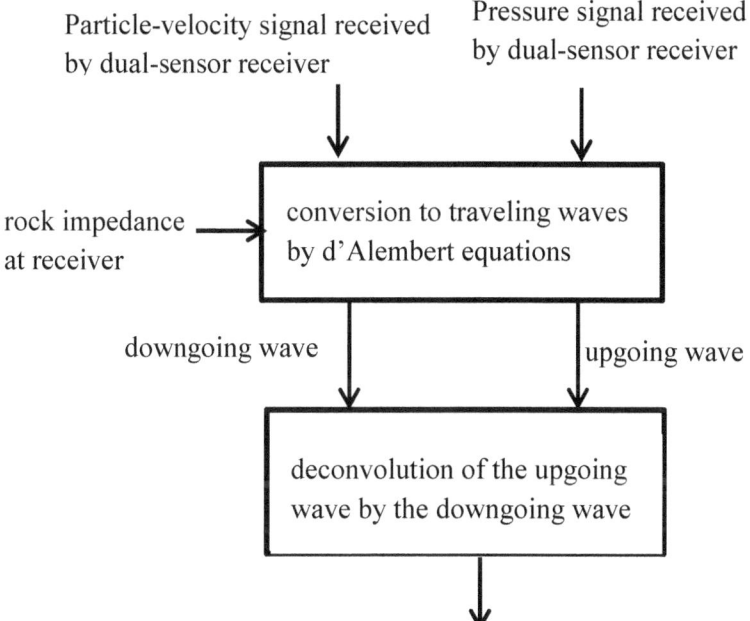

Fig. 5. Einstein deconvolution

The deconvolution of the upgoing by the downgoing wave at the buried receiver gives the unit-impulse reflection response of the subsystem below the receiver. This Einstein deconvolution process strips away the multiples and ghosts caused by upper system. It should be emphasized that the Einstein deconvolution process also strips away the unknown source signature wavelet. The resulting Einstein-deconvolved record is the unit-impulse reflection response of the geological interfaces below the receiver. Thus the output of the Einstein deconvolution process is precisely the input required for dynamic deconvolution. The output of dynamic deconvolution is the refection coefficient series. Thus Einstein deconvolution followed by dynamic deconvolution yields the series of reflection coefficients of the interfaces below the receiver. Instead of dynamic deconvolution, conventional ideconvolution can be used.

A receiver made up of dual geophone-hydrophone sensors measures two attributes of the wavefield. One attribute is particle velocity and the other attribute is pressure. The dual-sensor receiver is buried at a level below the level of the buried source. Einstein deconvolution, which requires the dual-sensor data, removes all the reverberations and ghosts due to interfaces above the receiver. Einstein deconvolution also removes the unknown source signature in the same operation. The resulting deconvolved seismogram is the unit-impulse reflection response that would be produced as if there were no layers at all above the buried receiver.

If desired, dynamic deconvolution can be performed on the unit-impulse reflection response obtained by Einstein deconvolution. The output of the dynamic deconvolution process is the sequence of reflection coefficients for the interfaces below the receiver. Einstein deconvolution operates under the same limitations as spiking deconvolution. As a result, the limitations for Einstein deconvolution can be addressed by the same means as is done for spiking deconvolution.

For any method of deconvolution, it is always assumed that the output signal is known. In general, deconvolution consists of two steps. The first step is to measure, estimate, or otherwise determine the input wavelet. The second step is to deconvolve the output signal by this input wavelet to yield an estimate of the unit-impulse response. In predictive (spiking) deconvolution, the input wavelet is estimated from knowledge of the output signal. In signature deconvolution, the input wavelet (the signature) is measured directly. Einstein deconvolution removes all effects introduced by everything above a given receiver location, see for example Loewenthal and Robinson (2000, *Geophysics*,

65, 293-303) and Robinson (1999). Einstein deconvolution uses a dual geophone/hydrophone sensor. The method can be described in two steps. The first step converts the particle velocity and pressure signals into downgoing wave and upgoing wave components at a receiver location. This step is carried out with d'Alembert's equations. Einstein deconvolution strips away the multiples and ghosts caused by the layering *above* a given receiver. It should be emphasized that Einstein deconvolution also strips away the unknown source signature wavelet. The resulting Einstein-deconvolved record is then an estimate of the unit-impulse reflection response of the geological layering situated *below* the given receiver.

Both deconvolution methods have much in common. The difference is in the fundamental assumptions that determine how the deconvolution operator is obtained. Two approximations are made in spiking deconvolution, namely the white approximation and the small approximation. By the white approximation we mean that the series of reflection coefficients (i.e., the reflectivity) is made up of a sample of uncorrelated random variables. In such a case the autocorrelation of the reflectivity is approximately a spike. By the small approximation we mean that the probability distribution of the reflectivity has a "small" standard deviation. The word small is used in a relative sense, depending upon the case at hand. For example, in some cases a standard deviation of 0.05 might be considered small; in other cases a standard deviation of 0.10. An autocorrelation coefficient is a sum of products. For this reason autocorrelation coefficients are second-order coefficients. Whiteness means that the autocorrelation coefficients, except for lag zero, are small. Smallness requires that correlation coefficients of higher order are also small. In other words, we worry about higher order correlation coefficients as well as the usual second order ones.

The common goal of both spiking deconvolution and Einstein deconvolution is to obtain the reflection coefficient series as the deconvolved signal. Both spiking deconvolution and Einstein deconvolution carry out the deconvolution process on the upgoing signal. Both spiking deconvolution and Einstein deconvolution have the same deconvolution operator, namely the inverse of the downgoing signal. Thus these two methods of deconvolution look like each other. The difference is in the fundamental assumptions that determine the way the deconvolution operator is obtained. The small white reflectivity hypothesis allows the spiking deconvolution operator to be computed by least-squares from the upgoing signal. In addition the small white reflectivity hypothesis eliminates to a large degree the necessity for the final dynamic-deconvolution step. Spiking deconvolution has the advantage of many year's usage. It is robust and stable in the presence of noise. Einstein deconvolution has the advantage the small white reflectivity hypothesis is not required. In this way Einstein deconvolution is more general. However Einstein deconvolution is more sensitive to noise. Ideally both methods can be used in conjunction with one another to obtain better results.

Chapter 8. Ghost reflections

Gauss: When a philosopher says something that is true then it is trivial. When he says something that is not trivial then it is false.

Summary

When both the source and the receiver are on the surface of a layered system, the source produces energy that travels downward and the receiver picks up reflected energy that travels upward. However, when both the source and the receiver are below the surface, the source produces energy that travels both downward and upward and the receiver picks up reflected energy that travels both upward and downward. Because the source provides energy, the source layer is active. The super-source system made up of the layers above the source is passive. Likewise the sub-source system made up of the layers below the source is passive. The net downgoing energy in any layer of a passive system is constant. The downgoing input from the source layer to the sub-source system consists of two components, the downgoing direct waves and the downgoing ghost waves. The downgoing source signature produces the direct waves. The upgoing source signature is reflected from the super-source system, thereby creating a secondary downgoing source of energy. This secondary source produces the ghost waves. The ghost waves mimic the direct waves but with delay. Thus each layer in the sub-source system has the same positive net downgoing energy, which favors the primary reflections that are reflected upward from the interfaces below the source. (The same reasoning shows that each layer in the super-source system has the same positive net upgoing energy, which favors the primary reflections that are reflected downward from the interfaces above the source.) It is the reflections from deep interfaces that are important in seismic imaging and hence the receiver should be placed in a layer within the sub-source system. The type of receiver used should be one that produces downgoing waves on one channel and upgoing waves on the other channel. Einstein deconvolution, namely the deconvolution of the upgoing waves by the downgoing waves, yields the unit-impulse reflection response of the layered system below the receiver. In this way the signature, the ghost reflections, and the entire surface and near-surface multiples from all of the interfaces above the receiver are eliminated. It follows that if surface multiples are a major problem in vibroseis prospecting, then the receivers, or at least some of them, should be buried below the near–surface layers.

Source ghost and receiver ghost

Einstein deconvolution is the deconvolution of the upgoing waves by the downgoing waves. Loewenthal and Robinson (2000, *Geophysics*, 65, 293-303) show that Einstein deconvolution yields the unit-impulse reflection response of the layered system below the receiver. Their work treats the case in which the receiver is buried but the source is on the surface. As a result no ghost reflections can occur. This chapter extends that work to the case of a buried source, so that ghost reflections are produced. It is shown that as long as the receiver is placed strictly below the source layer, then Einstein deconvolution still yields the unit-impulse reflection response of the layered system below the receiver. In other words, Einstein deconvolution removes the source signature, the ghost reflections, and all of the multiple reflections due to the interfaces above the receiver.

The existence of a strong reflecting interface above the seismic source causes a simple ghost reflection. Two strong reflecting interfaces bound the water layer, namely the water-surface and the water-bottom. These two interfaces produce reverberations in the water layer. These two concepts, namely ghosts and reverberations, are basic to geophysical analysis. This chapter considers the nature of more general ghost reflections and also the nature of more general reverberations. In fact, all the interfaces above the source produce ghost reflections. Also reverberations within any layer are due to the interplay between all of the interfaces above the layer and all of the interfaces below the layer.

Unless otherwise stated, the terms "ghost reflection" or "ghost" refers to a source-ghost reflection. See Fig. 1. Source-ghost reflections occur when the energy source is buried or submerged. Source-ghost reflections originate from energy that initially travels upward from the source. This upward energy is then reflected downward by a reflector above the source. Such a reflector can occur at the base of the weathering or at the surface of the ground or water. Source-ghost energy joins the downgoing wave train to change the effective waveshape. Source ghosts must be distinguished from receiver ghosts. Receiver-ghost reflections occur when the receiver is buried or submerged. See Fig. 2. Receiver-ghosts represent energy that is reflected from the surface and then picked up by the receiver. Receiver-ghost reflections are components of surface multiple reflections that occur on the seismogram.

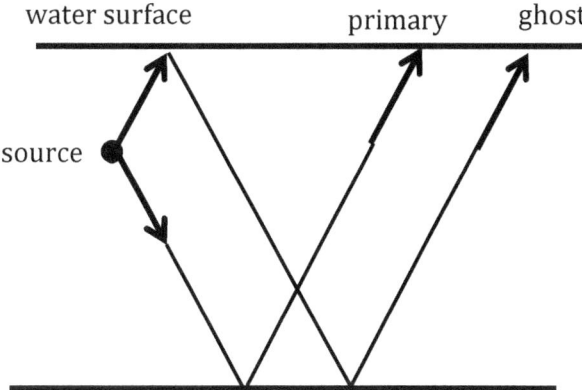

Fig. 1. The source ghost.

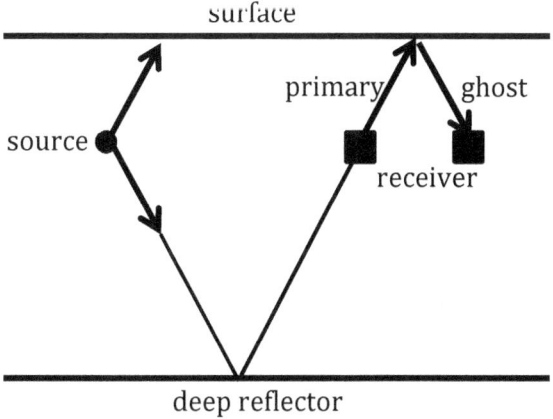

Fig. 2. The receiver ghost.

Consider the classic model upon which the seismic reflection method was originally based. The source was on the surface and so was the receiver. As a result all of the source energy was downgoing and the entire received signal was upgoing. In the model treated here the source is in a shot hole buried below the surface of the ground or, alternatively, the source is in the water below the surface. Furthermore the receiver is not on the surface, but is either buried in the earth or below the surface of the water. Thus in this model, both the source and the receiver are below the surface. One of three cases can occur: the receiver is above the source, the receiver is at the same level as the source, or the receiver is below the source. It is assumed that the receiver is a dual-sensor that picks up both pressure and particle-velocity waves, and then converts them to downgoing particle-velocity waves on one channel, and to upgoing particle-velocity waves on the other channel.

For the moment assume that the source produces an upgoing spike and a downgoing spike. In the convention used here, the positive direction of the vertical axis is downward, and the negative direction of the vertical axis is upward. Let the downgoing spike in particle velocity be denoted by δ^{down}. Particle velocity has a direction associated with it, so the upgoing spike in particle velocity has the opposite sign to the downgoing spike in particle velocity; that is $\delta^{up} = -\delta^{down}$. On the other hand, pressure is not a directional quantity, so the upgoing spike in pressure has the same sign as the downgoing spike in pressure. In summary, an ideal shot has this characteristic: the downgoing spike (which travels in the positive vertical direction) has positive particle velocity and positive pressure, and the upgoing spike (which travels in the negative vertical direction) has negative particle velocity but still a positive pressure. In actuality, a physical seismic source produces not a spike, but a signal of some duration called the source signature. Denote the downgoing source signature in particle velocity by the symbol S. The upgoing signature may have a different strength or shape than the downgoing signature. However it is assumed in this paper that there is no difference. Thus the upgoing source signature in particle velocity is $-S$. No assumption as to the phase characteristics of the source signature is made, so the source signature may or may not be minimum-delay. Furthermore it is assumed that the source signature is neither measured nor estimated. In other words, the source signature is completely unknown.

We use the classic model for a horizontally layered earth. The layered system with n interfaces numbered $1,2,3,\cdots,\alpha-1,\alpha,\alpha+1,\cdots,n$. Interface 1 is the surface of the ground or the surface of the water as the case may be. The source is in layer α, which is the layer between interface $\alpha-1$ and interface α. More specifically, the source is at the top of layer α, so the source is just below interface $\alpha-1$. Layer α is called the source layer. The receiver may be in any layer, which is called the receiver layer. The lowermost interface is interface n. All the material below interface n represents basement rock. No reflections occur in the basement.

System A (the super-source system) is defined as the system made up of interfaces above the source, namely interfaces $1,2,3,\cdots,\alpha-1$. System B (the sub-source system) is defined as the system made up of interfaces below the source, namely interfaces $\alpha,\alpha+1,\cdots,n$. Consider the downgoing source signature only. It can only produce primary reflections from System B. It cannot produce primary reflections from System A. Consider the upgoing source signature only. It can produce primary reflections from System A. It cannot produce primary reflections from System B. If the receiver is at the same level as the source, then all the primary reflections can be captured. If the receiver is at a different level, then no primaries can be captured for those interfaces between source and receiver.

There is little interest in the interfaces above the source, and thus the upgoing source signature produces primaries of little value. There is much interest in the deep interfaces occurring in System B, and thus the downgoing source signature produces a record of great value. One purpose of this chapter is to show how the recorded seismogram can be processed to yield the unit-impulse reflection response of the system below the receiver. The ideas required can be obtained by reference to the theory of ghost reflections and reverberations.

Ghost reflections and reverberations

Robinson (1966, Multichannel Z-transforms and minimum-delay, *Geophysics*, 31, 482-500) describes two geophysical problems in terms of the mathematics of the Z-transform. One is the problem of *seismic ghost reflections* and the other is the problem of *water-confined reverberations*. The methods are strictly valid only for flat horizontal interfaces and for vertical incidence. Only compressional waves are considered. The material is reviewed so it may be extended to more complex situations.

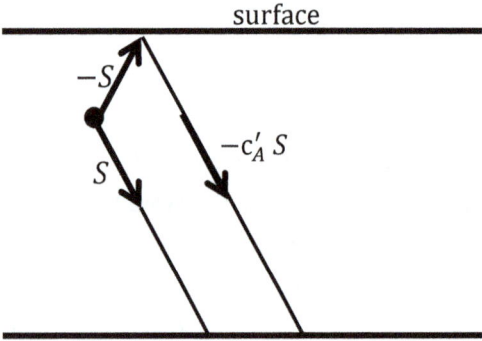

Fig. 3. The ghost-producing couplet in particle velocity

The recorded seismogram is actually the sum of two component seismograms, called the direct seismogram and the ghost seismogram. The explosion is set off in a shot hole drilled below the surface of the ground. The primary reflections are caused by the reflection from deep strata of the energy moving directly downward from the shotpoint. Meanwhile the energy that goes upward from the shotpoint is reflected from an overlying discontinuity, and thus there is a source of secondary energy moving directly downwards. The *ghost reflections* are caused by the reflection from deep strata of the energy moving directly downward from the source of secondary energy. Such a ghost reflection is not a primary reflection but instead it is the third leg of a multiple reflection. Thus a given deep reflecting horizon appears on the recorded seismogram as two reflection wavelets, displaced in time by twice the traveltime from the shot to the overlying discontinuity. Any differences in shape between the primary and ghost reflections can be attributed to various causes. However, in many instances the primary and ghost have approximately the same shape. In such a case, the two most important parameters become c'_A and n. The constant c'_A (which has magnitude less than unity) represents the Fresnel reflection coefficient of the overlying discontinuity for an upgoing incident wave. The constant n (assumed to be an integer) represents the time delay of the ghost with respect to the primary. The direct source (in particle velocity) going downward is S and the secondary source (in particle velocity) going downward is

$$-S\, c'_A\, Z^n \qquad (1)$$

The direct source produces the direct seismogram and the secondary source produces **the ghost seismogram.**

The *water reverberation* problem in marine seismic operations can be described in terms of the Z-transform. The water-air interface and the water-bottom interface are both strong reflectors. See Fig. 4.

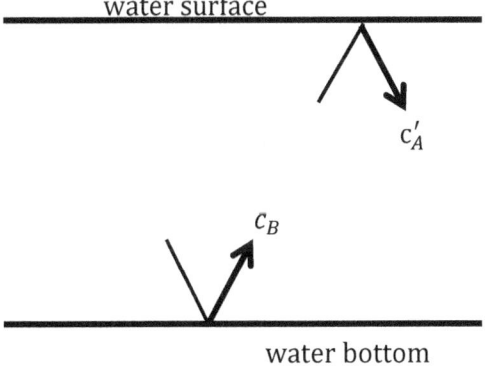

Fig. 4. The reflection coefficients involved in water reverberations

Because the water layer is a non-attenuating medium bounded by two strong reflecting interfaces, it represents an energy trap. A seismic pulse generated in this energy trap will be successively reflected between the two interfaces. Consequently, the water reverberation will obscure reflections from deep horizons below the water layer. The two-way traveltime in the water layer is n time units. A receiver that measures one-way-traveling-wave motion, both down and up, in the water layer is used. One channel records the downgoing reverberation in the water layer and the other channel, the upgoing reverberation in the water layer.

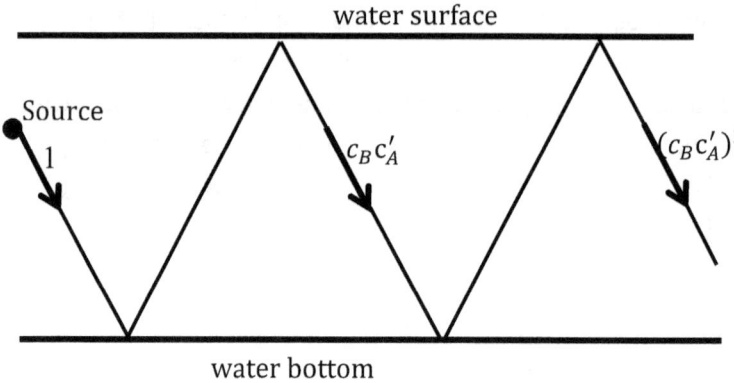

Fig. 5. A downgoing wave trapped in the water layer

Let c_B be the Fresnel reflection coefficient for a downgoing incident particle-velocity signal striking the water bottom, and let c'_A be the Fresnel reflection coefficient for an upgoing incident particle-velocity signal striking the water surface. Let the source produce a downgoing unit spike only. See Fig. 5. The resulting downgoing reverberation is made up of the following infinite series of components.

1. The initial downgoing source spike with amplitude 1 occurs at time 0.

2. The second downgoing spike with amplitude $c_B c_A'$ occurs at discrete time n This spike suffered a reflection at the bottom interface (reflection coefficient c_B) and a reflection at the top interface (reflection coefficient c_A').

3. The third downgoing spike with amplitude $(c_B c_A')^2$ occurs at discrete time $2n$. This spike suffered two reflections at the bottom and two reflections at the top.

4. The fourth downgoing spike with amplitude $(c_B c_A')^3$ occurs at discrete time $3n$. This spike suffered three reflections at the bottom and three reflections at the top.

Chapter 8. Ghost reflections

This sequence continues to infinity. The Z-transform of the water-confined downgoing reverberation spike-train is of the form

$$\Lambda = 1 + c_B c'_A Z^n + (c_B c'_A Z^n)^2 + (c_B c'_A Z^n)^3 + \cdots$$

This expression may be summed to give the following expression for the downgoing reverberation-producing filter

$$\Lambda = \frac{1}{1 - c_B c'_A Z^n} \tag{2}$$

Because $c'_A = -c_A$, equation (2) may also be written as

$$\Lambda = \frac{1}{1 + c_B c_A Z^n}$$

This filter is minimum-delay. Equation (2) also holds for an upgoing unit spike source. See Fig. 6. Equation (2) may be written as

$$\Lambda = 1 + c_B c'_A Z^n$$

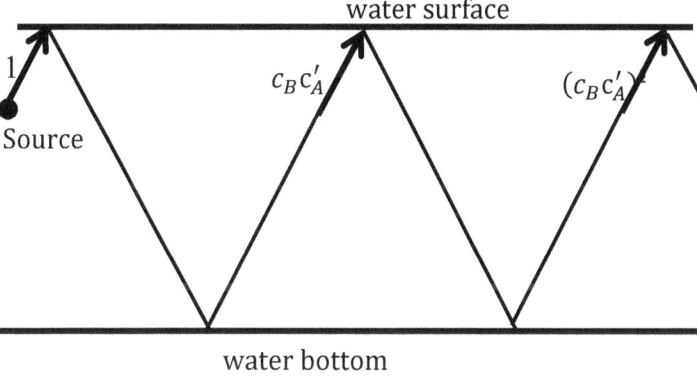

Fig. 6. An upgoing wave trapped in the water layer

System reflection and transmission coefficients

A Fresnel reflection or transmission coefficient is defined for just one interface. A system reflection coefficient and a system transmission coefficient are defined for a system of many layers. These system coefficients are, of course, functions of the Fresnel coefficients. Let the system have interfaces $1, 2, 3, \cdots, n$ with the respective Fresnel reflection coefficients $c_1, c_2, c_3, \cdots, c_n$. Let the source be a downgoing unit spike. The system reflection coefficient $R(Z)$ is the upgoing wave escaping into the air. The system transmission coefficient $T(Z)$ is the downgoing wave escaping into the basement.

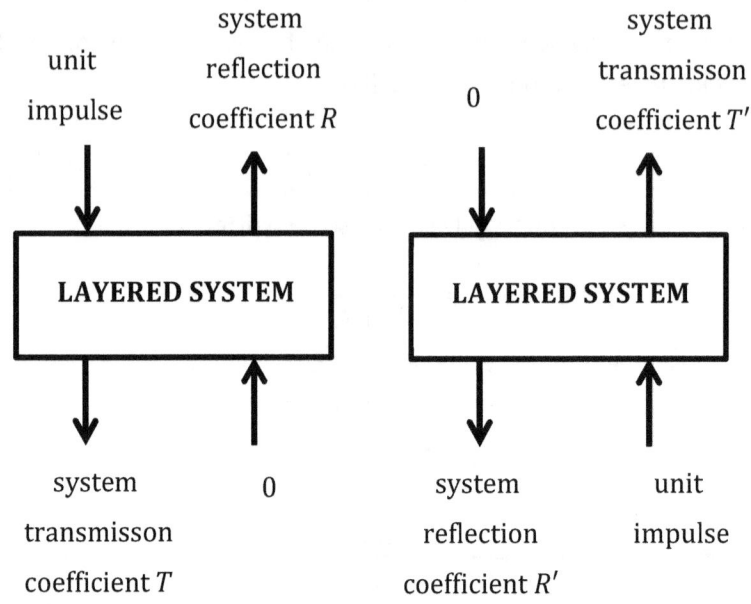

Fig. 7. The system coefficients

See Fig. 7 (Left). The system reflection coefficient and the system transmission coefficient are

$$R(Z) = \frac{-Q_n(Z)}{P_n(Z)} \quad \text{and} \quad T(Z) = \frac{\sigma_n \, Z^{n/2}}{P_n(Z)}$$

Here $P_n(Z)$ and $Q_n(Z)$ are the fundamental polynomials for the system and where σ_n is the one-way downgoing system transmission factor. The derivation is given in the Appendix at the end of this chapter.

See Fig. 7 (Right). Now let the source be an upgoing unit spike just below interface n. The system reflection coefficient $R'(Z)$ is the downgoing wave escaping into the basement, and the system transmission coefficient $T'(Z)$ is the upgoing wave escaping into the air. They are given by the equations

$$R'(Z) = \frac{Q_n^R(Z)}{P_n(Z)} \quad \text{and} \quad T'(Z) = \frac{\sigma_n' \, Z^{n/2}}{P_n(Z)}$$

Here $Q_n^R(Z)$ is the reverse of the polynomial $Q_n(Z)$ and σ_n' is the one-way upgoing system transmission factor. The system transmission coefficients T and T' are each minimum delay.

Ghost seismogram

Simple ghost reflections are always defined in the case of a single ghost-producing interface above the source. However the concept of a ghost can be generalized to the case of all the interfaces above the source. The Fresnel reflection coefficients of the system are given by $c_1, c_2, c_3, \cdots, c_n$. Break the given system into two component systems, denoted by A and B. System A, called the super-source system, contains all the interfaces above the source. The reflection coefficients of system A are $c_1, c_2, c_3, \cdots, c_{\alpha-1}$. System B, called the sub-source system, contains all the interfaces below the source. Their reflection coefficients are $c_\alpha, c_{\alpha+1}, \cdots, c_n$. In system A, the net energy flux is upward, whereas in system B the net energy flux is downward.

The fundamental polynomials can be found from the reflection coefficients for each system in isolation from the other systems. The fundamental polynomials for any system are identified by the subscript. For example, P_A and Q_A are the fundamental polynomials for the ghost-producing system A. Similarly P_B and Q_B are the fundamental polynomials for the sub-source system B. From the fundamental polynomials, the system reflection and transmission coefficients can be found. For example, we have

$$R_A'(Z) = \frac{Q_A^R(Z)}{P_A(Z)} \quad \text{and} \quad R_B(Z) = -\frac{Q_B(Z)}{P_B(Z)}$$

System A (the super-source system) has the system reflection coefficient R_A' for waves striking the system from below. System B (namely, the sub-source system) has the system reflection coefficient R_B for waves striking the system from above. The excited source produces a downgoing particle-velocity signature S as well as an upgoing particle-velocity signature $-S$. The ghost-producing secondary source can be obtained from equation (1) by replacing $c_A' Z^n$ by R_A'. Thus the ghost-producing secondary source is

$$-SR'_A$$

The source produces complex reverberations in the source layer, that is, in layer α. The mathematical structure is the same as given for the case of reverberations between two interfaces. However, now system coefficients must be used instead of Fresnel coefficients. Thus the system reflection coefficient R'_A is used for the layers above the source. Likewise, the system reflection coefficient R_B is used for the layers below the source. The reverberation-producing filter can be obtained from equation (2) by replacing $c_B \, c'_A \, Z^n$ by $R_B \, R'_A$. Thus the reverberation-producing filter is

$$\Lambda = \frac{1}{1 - R_B \, R'_A}$$

which gives

$$\Lambda = 1 + R_B \, R'_A \, \Lambda$$

It follows that the downgoing particle-velocity wave in the source layer for the direct seismogram is

$$D_\alpha^{\text{direct}} = S\Lambda$$

The downgoing particle-velocity wave in the source layer for the ghost seismogram is given by the reverberation

$$D_\alpha^{\text{ghost}} = -SR'_A \, \Lambda$$

The reflection of this downgoing particle-velocity wave from System B is the upgoing-reflected wave in the source layer. This wave for the direct seismogram is given by

$$U_\alpha^{\text{direct}} = R_B D_\alpha^{\text{direct}} = SR_B \Lambda$$

This wave for the ghost seismogram is given by

$$U_\alpha^{\text{ghost}} = R_B D_\alpha^{\text{ghost}} = -SR_B R'_A \, \Lambda$$

In addition, the upgoing signature in the source layer is

$$U_\alpha^{\text{signature}} = -S$$

The sum of the upgoing signature plus the upgoing ghost is

$$U_\alpha^{\text{signature}} + U_\alpha^{\text{ghost}} = -S - S \, R_B R'_A \, \Lambda = -S(1 + R_B R'_A \, \Lambda) = -S\Lambda$$

The total upgoing wave is

Chapter 8. Ghost reflections

$$U_\alpha = U_\alpha^{\text{signature}} + U_\alpha^{\text{ghost}} + U_\alpha^{\text{direct}} = -S\Lambda + SR_B\Lambda$$

The total downgoing wave is

$$D_\alpha = D_\alpha^{\text{direct}} + D_\alpha^{\text{ghost}} = S\Lambda - SR'_A\Lambda$$

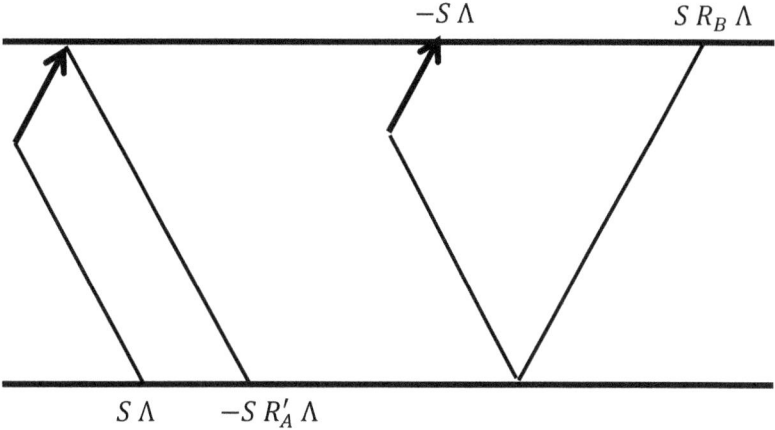

Fig. 8. (Left) Components of D_α. (Right) Components of U_α

Fig. 8 gives a schematic diagram illustrating the components of the equations. In the source layer the direct seismogram is made up of the downgoing reverberation $S\Lambda$ and the upgoing reverberation $SR_B\Lambda$. In the source layer the ghost seismogram is made up of the downgoing reverberation $-S\,R'_A\Lambda$ and the upgoing reverberation $-S\,R_B R'_A\Lambda$. In addition, there is the extraneous upgoing source signature $-S$.

Let the receiver is in layer β strictly below the source, so $\beta > \alpha$. Recall that system B contains all the interfaces below the source. System B which lies below the source has the series of reflection coefficients $c_\alpha, c_{\alpha+1}, \cdots, c_n$. The receiver can receive upgoing primary reflections for interfaces below the receiver. The receiver can also receive downgoing primary reflections from interfaces above the source. However, the receiver cannot receive primary reflections from interfaces between source and receiver. Let the downgoing and upgoing waves in the receiver layer be D_β and U_β respectively. Because system B is passive with a net downward energy flux these two waves are related by

$$U_\beta = R_F\, D_\beta$$

Here R_F is the reflection impulse response of the layers below the receiver. The above equation gives the equation for Einstein deconvolution as

$$R_F = \frac{U_\beta}{D_\beta}$$

Elimination of signature, ghosts, and near-surface multiples

The system between the receiver and the basement rock contains the reflection coefficients of interest in exploration. The near-surface reflection coefficients, that is, those of the interfaces above the receiver, give rise to the surface multiples and the ghosts that need to be eliminated. The dual channel one-way receiver measures the downgoing wave D_β and the upgoing wave U_β, so both are known. Einstein deconvolution is the deconvolution of the upgoing wave U_β by the downgoing wave D_β. The result is the unit-impulse reflection response R_F. That is, Einstein deconvolution gives the unit-impulse reflection response of the layered system below the receiver. This deconvolution process strips away the complex surface multiples and complex ghosts. It should be emphasized that this deconvolution process also strips away the unknown source signature S.

A buried source produces upgoing energy and downgoing energy. The direct seismogram results from the downgoing energy. The reflection of the upgoing energy from the overlying interfaces produces a secondary source of downgoing energy. The ghost seismogram results from this secondary source of downgoing energy. A buried source is in effect bounded, on the one hand, by all the interfaces above the source and, on the other hand, by all the interfaces below the source. These bounds produce reverberations in the source layer.

The dual channel receiver can be used to receive both the pressure wave and the particle velocity wave. From these two waves, both the downgoing wave and the upgoing wave at the receiver bassn be computed. Einstein deconvolution, which is the deconvolution of the upgoing wave by the downgoing wave, removes the source signature, the ghost reflections, and the near-surface multiple reflections. The deconvolved record is the unit-impulse reflection response of the layers below the receiver.

A dual channel one-way receiver buried at depth below the source receives both the downgoing and the upgoing waves. Deconvolution of the upgoing wave by the downgoing wave removes the (unknown and not necessarily minimum-delay) source signature as well all the surface multiples and all the ghosts. The resulting deconvolved record is the unit-impulse reflection response of the geological interfaces below the receiver.

Appendix

Interface 1 is the surface. The datum is a fictitious interface one-half time unit above the surface. The reflection coefficient for the datum is zero. Let the source be a downgoing unit spike at the datum. The resulting upgoing escaping wave at the datum is the system reflection coefficient $R(Z)$. The resulting downgoing escaping wave into the basement is the system transmission coefficient $T(Z)$. The one-way downgoing system transmission factor and the one-way upgoing system transmission factor are defined respectively as

$$\sigma_n = (1 + c_1)(1 + c_2) \cdots (1 + c_n)$$
$$\sigma'_n = (1 - c_1)(1 - c_2) \cdots (1 - c_n)$$

First consider a system made up of a single interface with reflection coefficient c_1. For a downward unit spike incident on the interface, the transmission response is $1 + c_1$ and the reflection response is c_1. These responses can be written as

$$T_1 = \frac{(1 + c_1) Z^{1/2}}{1} \quad \text{and} \quad R_1 = -\frac{-c_1 Z}{1}$$

Define the fundamental polynomials as

$$P_1(Z) = 1 \quad \text{and} \quad Q_1(Z) = -c_1 Z$$

So

$$T_1 = \frac{(1 + c_1) Z^{1/2}}{P_1(Z)} \quad \text{and} \quad R_1 = -\frac{Q_1(Z)}{P_1(Z)}$$

Define the reverse polynomials (with the superscript R denoting reverse) as

$$P_n^R(Z) = Z^n P_n(Z^{-1}) \quad \text{and} \quad Q_n^R(Z) = Z^n Q_n(Z^{-1})$$

Because technically the fundamental polynomials for $n = 1$, are defined as polynomials of degree 1, we may write these polynomials as

$$P_1(Z) = 1 + 0\, Z \quad \text{and} \quad Q_1(Z) = 0 - c_1 Z$$

This allows us to write the corresponding reverse polynomials as

$$P_1^R(Z) = 0 + 1\, Z \quad \text{and} \quad Q_1^R(Z) = -c_1 + 0\, Z$$

or

$$P_1^R(Z) = Z \quad \text{and} \quad Q_1^R(Z) = -c_1$$

Next consider a system made up of a two interfaces with reflection coefficients c_1 and c_2. For a downward unit spike incident on the interface, the transmission and reflection responses are

$$T_2 = (1 + c_1)(1 + c_2)\Lambda\, Z^{1/2} \quad \text{and} \quad R_2 = c_1 Z + (1 - c_1^2) c_2 \Lambda\, Z^2$$

where the reverberation operator is

$$\Lambda = \frac{1}{1 + c_1 c_2 Z}$$

Thus the responses are

$$T_2 = \frac{(1 + c_1)(1 + c_2)\, Z^{1/2}}{1 + c_1 c_2 Z} \quad \text{and} \quad R_2 = -\frac{-c_1 Z - c_2 Z^2}{1 + c_1 c_2 Z}$$

Define the fundamental polynomials as

$$P_2(Z) = 1 + c_1 c_2 Z \quad \text{and} \quad Q_2(Z) = -c_1 Z - c_2 Z^2$$

Thus

$$T_2 = \frac{(1 + c_1)(1 + c_2)\, Z}{P_2(Z)} \quad \text{and} \quad R_2 = -\frac{Q_2(Z)}{P_2(Z)}$$

Because

$$P_1(Z) = 1 \quad \text{and} \quad Q_1(Z) = -c_1 Z$$

$$P_1^R(Z) = Z \quad \text{and} \quad Q_1^R(Z) = -c_1$$

It follows that

Chapter 8. Ghost reflections

$$P_2(Z) = 1 - c_2 Z(-c_1)$$
$$Q_2(Z) = -c_1 Z - c_2 Z(Z)$$

or

$$P_2(Z) = P_1(Z) - c_2 Z Q_1^R(Z)$$
$$Q_2(Z) = Q_1(Z) - c_2 Z P_1^R(Z)$$

It can be shown that the following recursion holds to generate the polynomials for an arbitrary number of interfaces

$$P_n(Z) = P_{n-1}(Z) - c_n Z Q_{n-1}^R(Z)$$
$$Q_n(Z) = Q_{n-1}(Z) - c_n Z P_{n-1}^R(Z)$$

and the responses are

$$T_n = \frac{\sigma_n Z^{n/2}}{P_n(Z)} \quad \text{and} \quad R_n = -\frac{Q_n(Z)}{P_n(Z)}$$

The polynomial $P_n(Z)$ is minimum delay. The polynomials $P_n(Z)$ and $Q_n^R(Z)$ are each of degree $n-1$ whereas the polynomials $Q_n(Z)$ and $P_n^R(Z)$ are each of degree n.

Now let the source be an upgoing unit spike just below interface n. The system reflection coefficient $R'(Z)$ is the resulting downgoing wave just below interface n, and the system transmission coefficient $T'(Z)$ is the resulting upgoing wave at the datum. The upgoing system coefficients as given by

$$T'_n = \frac{\sigma'_n Z^{n/2}}{P_n(Z)} \quad \text{and} \quad R'_n = \frac{Q_n^R(Z)}{P_n(Z)}$$

Because a water reverberation involves the water surface, it is one type of surface multiple reflection. However, many other types exist. They involve the water surface and various interfaces at depth. The purpose of this paper is to show how the source signature, the ghosts, and all the near-surface multiples can all be removed by Einstein deconvolution. The ghosts are complex ghosts-produced by layered system that includes all the interfaces above the source. The surface multiples are complex surface multiples produced by the interplay of the layers above the receiver with the layers below the receiver.

Chapter 9. Fourier series and Fourier transform

Gauss: Further, the dignity of the science itself seems to require that every possible means be explored for the solution of a problem so elegant and so celebrated.

Periodic functions

A periodic function is one which repeats itself indefinitely. Let t denote time. A function $f(t)$ is called periodic if there is a positive constant T for which

$$f(t + T) = f(t)$$

The constant T is called the period. It is easily verified that the sum, difference, product, or quotient of two periodic functions with the same period T is again a periodic function of period T.

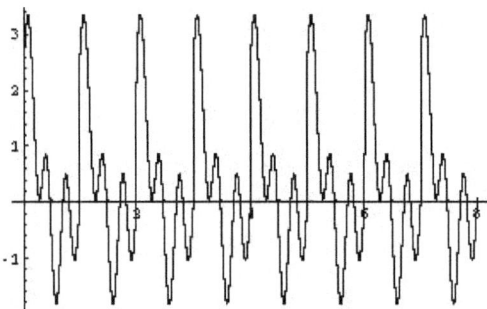

Fig. 1. Periodic function

If a periodic function is plotted on a closed interval with length equal to the period, then the entire graph of the function can be obtained by periodic repetition of the portion of the graph corresponding to this interval. This property is illustrated in Fig. 1.

For $e^{i\omega t}$ to be periodic with period T, it is required that

$$e^{i\omega(t+T)} = e^{i\omega t} = e^{i\omega t} e^{i\omega T}$$

This implies that

$$e^{i\omega T} = 1$$

Therefore

$$\omega T = 2\pi n \quad \text{where} \quad n = 0, \pm 1, \pm 2, \cdots$$

As a result each of the frequencies $\omega = 2\pi n/T$ satisfies the constraint

$$f(t+T) = f(t)$$

The fundamental frequency, which is given by $n = 1$, is $\omega = 2\pi/T$. The fundamental period is $T = 2\pi/\omega$. It follows that T is also the fundamental period of $\cos(\omega t)$ and $\sin(\omega t)$.

A type of motion that is very frequently observed in nature is vibratory motion; for example, (a) the back-and-forth movement of a pendulum, (b) the up-and-down motion of a bobbing spring, and (c) the oscillations of the prongs of a tuning fork. In fact, vibratory motion is typical of all waves, whether seismic waves, water waves, visible light, radio waves, x-rays, or gamma-radiation. The simplest form of vibratory motion is sinusoidal motion, also known as simple harmonic motion. The equation is

$$f(t) = A \sin(\omega t + \varphi)$$

This function is called a sinusoid of amplitude $|A|$ angular frequency ω, and initial phase φ. As we have just seen, the fundamental period of the sinusoid is $T = 2\pi/\omega$.

Why is the use of sines and cosines of the time variable t appropriate for situations that are relevant to much of theory of wave motion? The reason is that sines and cosines let us define frequency, and that the idea of frequency is basic to digital filtering. The cyclical frequency is customarily denoted by the symbol f. However, the use of the symbol f for frequency should not be confused with the standard use of the same symbol for function.

What about negative frequencies? Are they encountered in seismic work? In seismic work, only real-valued signals are recorded. As a result, a positive frequency cannot be distinguished from its negative. For example, for a real-valued signal a frequency of 10 Hz is indistinguishable from a frequency of -10 Hz. Hence only positive frequencies need be considered in ordinary circumstances.

Period is a general word that is used to denote an interval of time, but the word period has special meaning in frequency analysis. By the *period* of a repeating process, we mean the time-interval during which the process exactly repeats itself.

What does period have to do with the subject of frequency? How are they related? The period T and the frequency f are reciprocals of each other, that is,

$$T = \frac{1}{f} \quad \text{and} \quad f = \frac{1}{T}$$

The quantity $1/T$ is the number of oscillations in an interval T, thus explaining the term frequency. The dimension of f is cycles/second.. The dimension of T is seconds per cycle. seconds cycles/. period is the number of seconds per cycle. As we have seen, hertz is a measure of cyclic frequency. Hertz denotes the number of cycles per second.

The other type of frequency is angular frequency ω, which is measured in radians per second. Angular frequency is related to the cyclic frequency by the fundamental relation $\omega = 2\pi f$. The factor 2π in this equation comes from the fact that there are 2π radians (or 360°) in each cycle. The units of ω are (radians/second.

In general usage, the word frequency can be used for either ω or f, so one must determine from the context which one is being referred to. However, when the word frequency is used alone, it usually refers to cyclic frequency in Hertz.

The sinusoid

$$y = \cos\left(3t + \frac{\pi}{3}\right)$$

has period $2\pi/3$ and initial phase $\pi/3$. From trigonometry, we know

$$A\cos(\omega t + \varphi) = A(\cos \omega t \cos \varphi - \sin \omega t \sin \varphi)$$

If we set

$$a = A\cos\varphi \quad \text{and} \quad b = -A\sin\varphi$$

then

$$A\cos(\omega t + \varphi) = a\cos\varphi + b\sin\varphi$$

Often we shall write sinusoids in this form. For example, we have

$$\cos\left(3t + \frac{\pi}{3}\right) = \frac{1}{2}\cos 3t + \frac{\sqrt{3}}{2}\sin 3t$$

Jean Baptiste Joseph Fourier

Jean Baptiste Joseph, Baron de Fourier (1768 - 1830) was a friend of Napoleon and accompanied his master to Egypt in 1798, where he developed an interest in hieroglyphics. Upon his return, he became prefect of the district of Isere in southeastern France, and in this capacity had the first modern road built from Grenoble to Turin. He also befriended the boy, Jean François Champollion (1790-1832), who later deciphered the Rosetta stone. During these years, Fourier worked on the theory of the conduction of heat which led to his use of infinite series in terms of sines and cosines.

The problem of what functions can be represented by sinusoidal series arose in Fourier's researches. In connection with the study of the problem of heat transfer, he was confronted with the following problem. Let the given function be the discontinuous step function

$$h(x) = \begin{cases} -1, & -\pi < x < 0 \\ 1, & 0 < x < \pi \end{cases}$$

Fourier said that the function could be expanded into a trigonometric series, namely the Fourier sine series given by

$$h(x) = \sum_{k=1}^{\infty} b_k \sin kx, \quad -\pi < x < 0 \qquad (1)$$

The function is an odd function, that is, $h(x) = -h(x)$. Fourier developed a method to determine the Fourier coefficients b_k. The method is equivalent to a method originally given by Euler. The required expression for the coefficient (called the Fourier coefficient) is given by

$$b_n = \frac{1}{\pi}\int_{-\pi}^{\pi} h(x)\sin nx\, dx = \frac{2}{\pi}\int_{0}^{\pi} h(x)\sin nx\, dx \qquad (2)$$

We now substitute the expression for $h(x)$ given by (1) into equation (2). We obtain

$$b_n = \frac{1}{\pi} \int_{-\pi}^{0} (-1) \sin nx \, dx + \frac{1}{\pi} \int_{0}^{\pi} (1) \sin nx \, dx$$

But

$$-\int_{-\pi}^{0} \sin nx \, dx = \left[\frac{\cos nx}{n}\right]_{x=-\pi}^{0} = \frac{1}{n} - \frac{(-1)^n}{n} = \begin{cases} \frac{2}{n}, & n = 1,3,5,\cdots \\ 0, & n = 2,4,6,\cdots \end{cases}$$

Also

$$\int_{0}^{\pi} \sin nx \, dx = -\left[\frac{\cos nx}{n}\right]_{x=0}^{\pi} = \frac{1}{n} - \frac{(-1)^n}{n} = \begin{cases} \frac{2}{n}, & n = 1,3,5,\cdots \\ 0, & n = 2,4,6,\cdots \end{cases}$$

Therefore the Fourier coefficients are

$$b_n = \begin{cases} \frac{4}{\pi n}, & n = 1,3,5,\cdots \\ 0, & n = 2,4,6,\cdots \end{cases}$$

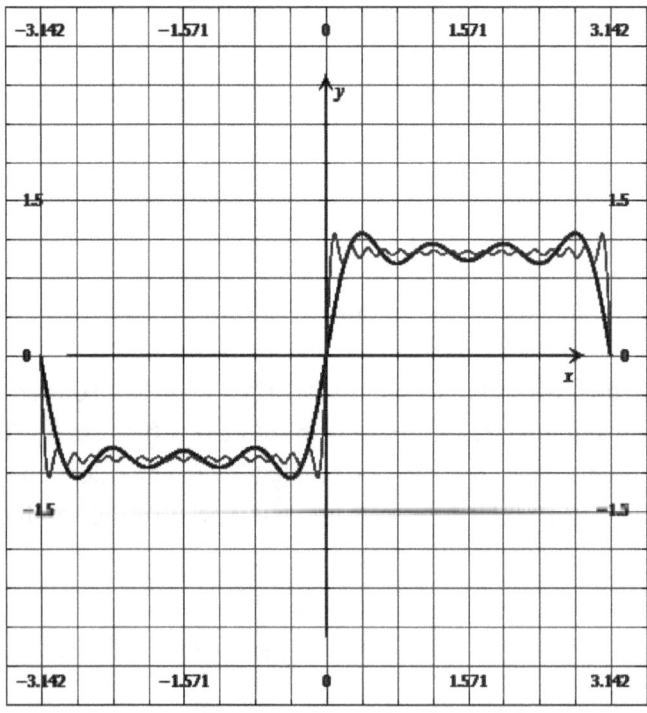

Fig. 2. The heavy curve is the 4 term approximation. The light curve is the 17 term approximation.

Chapter 9. Fourier series and Fourier transform

Thus Fourier showed that equation (1) becomes the infinite series

$$h(x) = \frac{4}{\pi}\left(\frac{\sin x}{1} + \frac{\sin 3x}{3} + \frac{\sin 5x}{5} + \frac{\sin 7x}{7} + \cdots\right)$$

Let us approximate $h(x)$ by the finite series of 4 terms given by

$$h(x) \approx \frac{4}{\pi}\left(\frac{\sin x}{1} + \frac{\sin 3x}{3} + \frac{\sin 5x}{5} + \frac{\sin 7x}{7}\right)$$

Also let us approximate $h(x)$ by the finite series of 17 terms given by

$$h(x) \approx \frac{4}{\pi}\left(\frac{\sin x}{1} + \frac{\sin 3x}{3} + \frac{\sin 5x}{5} + \cdots + \frac{\sin 33x}{33}\right)$$

The two graphs are shown in Fig. 2. The 17 term approximation fits the function $h(x)$ much better than the 4 term approximation. The figure illustrates (1) that finite-term approximations overshoot at a jump discontinuity, and (2) that this overshoot does not die out as the number of terms increases. This important phenomenon is known as the Gibbs effect.

Fourier did not prove that the trigonometric series (1) does converge to the function $h(x)$, but this question was answered in the affirmative by later investigations. In any case, it is important that Fourier first showed that it was possible for a trigonometric series to represent a given discontinuous function. It is an entirely different question to find the general mathematical conditions under which this series does indeed converge and does really possess this function as its sum.

In a session of the French Academy on December 21, 1807, Joseph Fourier announced his result that the sinusoidal series (15) represents the discontinuous function (8). This result inaugurated a new chapter in the development of mathematics. The older members of the Academy, including Lagrange, regarded the claim of Fourier as entirely incredible. In fact, Fourier asserted that any arbitrary function, defined in a finite interval by an arbitrary graph, can always be represented as a sum of pure sine and cosine functions. The reason why the academicians questioned Fourier's theorem was that they believed that any superposition of sine and cosine functions could never give anything but an infinitely differentiable function which we call "analytic". An analytic function is very far from some arbitrarily drawn graph with kinks and discontinuities. In fact, an analytic function has the property that, given its values in an arbitrarily small interval, the continuation of its values to the right and left is uniquely determined by expansions in Taylor series. How can this property be reconciled with the generality claimed by Fourier's theorem?

A century of intense mathematical research, which culminated in the work of Henri Lebesgue in 1904, demonstrated that Fourier's claim was entirely justified. Fourier's discovery has become the cornerstone of many fundamental investigations in pure and applied mathematics. The possibility of resolving all of the ordinarily occurring functions in the physical universe into pure sine and cosine vibrations has produced profound results in the physical and engineering sciences. We will now state Fourier's theorem: Any arbitrary (integrable) function on a finite interval can be represented by a Fourier series.

The concept of analytic function requires a strong interconnection of the values of a function, where knowledge at one point allows us to predict the value at a point at a finite distance h. This prediction mechanism is embodied in the Taylor series expansion. However, a nonanalytic function, such as a rough and discontinuous function, does not demand any such prediction mechanism between the immediate vicinity of a point and its wider surroundings. The Fourier series expansion is stated in terms of this wider concept of function. The coefficients of a Fourier series, as shown by the Euler-Lagrange result, are obtained by integration and not, as in the case of Taylor series, by differentiation. Each Fourier coefficient b_n is obtained by integrating $h(x) \sin nx$ over the entire range. Thus, any modification of $h(x)$ in a limited portion of the range changes all of the Fourier coefficients. It follows that the interconnections operate in the Fourier series in a global sense and not in a local sense as in the case of the Taylor series. It is the behavior of $h(x)$ in the large that matters in the case of the Fourier series, and not so much the behavior in the vicinity of a point. How can we resolve the differences between these two types of expansions? They seem very different at first glance. The Taylor series, which is the expansion about a point, gives strict predictions a finite distance from the point; the Fourier series, which is an expansion in the large, gives knowledge of the function in the entire range. The Taylor series requires unlimited differentiability at a point; the Fourier series does not demand any differentiability properties whatsoever. Surprisingly enough, the chasm between the Taylor series and the Fourier series is bridged by means of the Z-transform, which is the fundamental transform used in the theory of digital signal processing.

General form of Fourier series

If a function $f(x)$ satisfies $f(x) = f(2\pi + x)$ then the function is said to have period 2π. Such a function $f(x)$ can be represented by a trigonometric Fourier series of the form

$$f(x) = \frac{a_0}{2} + \sum_{n=1}^{\infty} a_n \cos nx + \sum_{n=1}^{\infty} b_n \sin nx$$

and the Fourier coefficients are

$$a_0 = \frac{1}{\pi} \int_{-\pi}^{\pi} f(x) dx$$

$$a_n = \frac{1}{\pi} \int_{-\pi}^{\pi} f(x) \cos nx \, dx, \qquad n = 1,2,3,\cdots$$

$$b_n = \frac{1}{\pi} \int_{-\pi}^{\pi} f(x) \sin nx \, dx, \qquad n = 1,2,3,\cdots$$

A function $f(x)$ is said to be even if $f(-x) = f(x)$ for all x and odd if $f(-x) = -f(x)$ for all x. Show that if $f(x)$ is an odd function then

$$\int_{-\pi}^{\pi} f(x) dx = 0$$

If $f(x)$ is an odd function, then all the Fourier coefficients a_n for $n = 0,1,2,3,\cdots$ are zero. In such a case we have the Fourier sine series

$$f(x) = \sum_{n=1}^{\infty} b_n \sin nx$$

If $f(x)$ is an even function, then

$$\int_{-\pi}^{\pi} f(x) dx = 2 \int_{0}^{\pi} f(x) dx$$

If $f(x)$ is an even function, then all the Fourier coefficients b_n for $n = 1,2,3,\cdots$ are zero. In such a case we have the Fourier cosine series

$$f(x) = \frac{a_0}{2} + \sum_{n=1}^{\infty} a_n \cos nx$$

$$f(x) = a_0 + \sum_{n=1}^{\infty} a_n \cos nx$$

For any function $f(x)$ with arbitrary period T, the change of variables $x = 2\pi t/T = \omega t$ can be used to transform the interval of integration from $[-\pi, \pi]$ to $[-T/2, T/2]$. The angular frequency is $2\pi/T = \omega$. The Fourier series is

$$f(t) = \frac{a_0}{2} + \sum_{n=1}^{\infty} a_n \cos n\omega t + \sum_{n=1}^{\infty} b_n \sin n\omega t$$

and the Fourier coefficients are

$$a_0 = \frac{2}{T} \int_{-T/2}^{T/2} f(t) dt$$

$$a_n = \frac{2}{T} \int_{-T/2}^{T/2} f(t) \cos n\omega t\, dt, \qquad n = 1,2,3,\cdots$$

$$b_n = \frac{2}{T} \int_{-T/2}^{T/2} f(t) \sin n\omega t\, dt, \qquad n = 1,2,3,\cdots$$

The product of two even functions is even, the product of two odd functions is even and the product of an even and an odd function is odd.

Complex Fourier series

We will use the trigonometric formulas

$$\cos n\omega t = \frac{(e^{in\omega t} + e^{-in\omega t})}{2}$$

$$\sin n\omega t = \frac{(e^{in\omega t} - e^{-in\omega t})}{2}$$

Thus the Fourier series becomes

Chapter 9. Fourier series and Fourier transform

$$f(t) = \frac{a_0}{2} + \frac{1}{2}\sum_{n=1}^{\infty} a_n\left(e^{in\omega t} + e^{-in\omega t}\right) + \frac{1}{2i}\sum_{n=1}^{\infty} b_n\left(e^{in\omega t} - e^{-in\omega t}\right)$$

The above equation becomes

$$f(t) = \frac{a_0}{2} + \frac{1}{2}\sum_{n=1}^{\infty} \frac{a_n - ib_n}{2} e^{in\omega t} + \frac{1}{2i}\sum_{n=1}^{\infty} b_n \frac{a_n + ib_n}{2} e^{-in\omega t}$$

If we define the complex coefficients

$$c_0 = \frac{a_0}{2}, \quad c_n = \frac{a_n - ib_n}{2}, \quad c_{-n} = \frac{a_n + ib_n}{2}$$

then the Fourier series

$$f(t) = \frac{a_0}{2} + \frac{1}{2}\sum_{n=1}^{\infty} a_n\left(e^{in\omega t} + e^{-in\omega t}\right) + \frac{1}{2i}\sum_{n=1}^{\infty} b_n\left(e^{in\omega t} - e^{-in\omega t}\right)$$

becomes

$$f(t) = c_0 + \sum_{n=1}^{\infty} c_n e^{in\omega t} + \sum_{n=1}^{\infty} c_{-n} e^{-in\omega t} = \sum_{n=-\infty}^{\infty} c_n e^{in\omega t}$$

Thus the complex Fourier series is

$$f(t) = \sum_{n=-\infty}^{\infty} c_n e^{in\omega t} \quad \text{where} \quad c_n = \frac{2}{T}\int_{-T/2}^{T/2} f(t) e^{-in\omega t} dt$$

Fourier transform

The Fourier transform applies to non-periodic functions that meet certain restrictions. Essentially the period T is allowed to get infinitely large. Instead of a discrete sequence of coefficients c_n, we now have a continuous function $F(\omega)$. Thus the Fourier transform is

$$f(t) = \frac{1}{2\pi}\int_{-T/2}^{T/2} F(\omega) e^{-i\omega t} d\omega \quad \text{where} \quad F(\omega) = \int_{-\infty}^{\infty} f(t) e^{i\omega t} dt$$

The function $F(\omega)$ is called as the Fourier transform of f(t), and the equation form f(t) is called the inverse Fourier Transform.

Discrete Fourier transform (DFT)

Because digital computers are used to perform Fourier analysis, it is necessary to require that both the time and the frequency variables are discrete. The discrete Fourier Transform (DFT) is a numerical approximation to the Fourier transform. The following steps outline how to convert the Fourier Transform (FT) into the Discrete Fourier Transform:

1) Assume the sampling window is T. The number of sampling points is N.

2) Define the sampling increment $\Delta T = T_s = T/N$. Define the sample points $t_k = k(\Delta T)$ for $k = 0, \ldots, (N-1)$.

3) Define the signal values at each sampling points as $f_k = f(t_k)$.

4) Define the frequency sampling points $\omega_n = 2\pi n/T$, where $2\pi n/T$ is termed as the fundamental frequency.

5) Consider the problem of approximating the Fourier transform of $f(t)$ at the points $\omega_n = 2\pi n/T$. The answer is

$$F(\omega_n) = \int_{-\infty}^{\infty} f(t) e^{-i\omega t} dt \quad \text{for} \quad n = 0, 1, \cdots, N-1$$

Approximate the above equation by

$$F(\omega_n) = \sum_{k=0}^{N-1} f(t_k) e^{-i\omega t_k} \quad \text{for} \quad n = 0, 1, \cdots, N-1$$

This is the Discrete Fourier Transform. The inverse Discrete Fourier Transform is defined as

$$f(t_k) = \frac{1}{N} \sum_{k=0}^{N-1} F(\omega_n) e^{i\omega t_k} \quad \text{for} \quad n = 0, 1, \cdots, N-1$$

Fast Fourier Transform (FFT) is an effective algorithm of Discrete Fourier Transform (DFT). This algorithm reduces the computation time of DFT for N points from N^2 to $N \log_2(N)$. The only requirement of this algorithm is that number of point in the series have to be a power of 2 (2^n points) such as 32, 1024, 4096. Zero padding is used at the end of the data set if the sampling number is not equal to the exact the power of 2.

The total sampling period is T. The base frequency $2/T$ represents the lowest frequency of the signal we can see in the frequency domain. On the other hand, the Nyquist frequency

$$f_{\text{Nyquist}} = \frac{1}{2\,\Delta t}$$

represents the highest frequency of the signal that we can see in the frequency domain.

Exercises

1. Many functions can be written as a power series. An example is provided by the geometric series:

$$\frac{1}{1-x} = 1 + x + x^2 + x^3 + x^4 + \cdots$$

which is valid for $-1 < x < 1$.

2. Find the Taylor series for $f(x) = e^x$ with center $x_0 = 0$. All derivatives are of the form e^x, so at the center they all evaluate to 1. Thus the Taylor series is

$$e^x = 1 + x + \frac{x^2}{2!} + \frac{x^3}{3!} + \frac{x^4}{4!} + \cdots$$

3. Find the Fourier series of the function of period 2π defined by

$$f(x) = \begin{cases} -1, & x < 0 \\ 1, & x \geq 0 \end{cases}$$

f(x) = -1 if -π < x < 0, and f(x) = 1 if 0 < x < π. What does the Fourier series converge to at x = 0? Answer:

$$f(x) = \frac{4}{\pi} \sum_{n=0}^{\infty} \frac{\sin(2n+1)x}{(2n+1)}$$

4. What is the Fourier series of the function of period 2π defined by

$$f(x) = \begin{cases} 1, & -\pi < x < 0 \\ 3, & 0 < x < \pi \end{cases}$$

Answer:

$$f(x) = 2 + \frac{4}{\pi} \sum_{n=0}^{\infty} \frac{\sin(2n+1)x}{(2n+1)}$$

5. Find the Fourier series of $f(x) = x^2$ where $0 < x < 2p$ and $f(x)$ has period $2p$. Answer:

$$f(x) = \frac{4\pi^2}{3} + 4\sum_{n=1}^{\infty} \frac{\cos nx}{n^2} - 4\pi \sum_{n=1}^{\infty} \frac{\sin nx}{n}$$

6. Find the Fourier series of the function $f(x) = x, -\pi < x < \pi$

Answer. It is an odd function. In such a case we have a Fourier sine series, with coefficients

$$b_n = \frac{1}{\pi}\int_{-\pi}^{\pi} x \sin nx \, dx = \frac{1}{\pi}\left[-\frac{x\cos nx}{n} + \frac{\sin nx}{n^2}\right]_{-\pi}^{\pi}$$

Thus

$$b_n = -\frac{2}{n}\cos n\pi = \frac{2}{n}(-1)^{n+1}$$

so the Fourier sine series is

$$f(x) = 2\left(\sin x - \frac{\sin 2x}{2} + \frac{\sin 3x}{3} - \cdots\right)$$

7. Find the Fourier series of the function

$$f(x) = \begin{cases} 0, & -\pi \le x < 0 \\ \pi, & 0 \le x \le \pi \end{cases}$$

Answer.

$$f(x) = \frac{\pi}{2} + 2\left(\sin x + \frac{\sin 3x}{3} + \frac{\sin 5x}{5} + \cdots\right)$$

Chapter 10. Gauss and Maxwell's equations

Faraday, Ampere, and Gauss equations

Maxwell undertook the problem of solving the relationship between electric currents and magnetic fields. In this section we will derive Maxwell's equations. They involve the vector quantities

$$E = \text{electic intensity}$$
$$H = \text{magnetic intensity}$$
$$D = \varepsilon E = \text{electric flux density}$$
$$B = \mu H = \text{magnetic flux density}$$
$$J = \text{current density}$$

and the scalars

$$\varepsilon = \text{permittivity}$$
$$\mu = \text{permeability}$$
$$\sigma = \text{conductivity}$$
$$\rho = \text{charge density}$$

$$q = \iiint_V \rho \, dV = \text{total charge within volume } V \quad (1)$$

$$\varphi = \iint_S N \cdot B \, dS = \text{total magnetic flux through surface } S \quad (2)$$

$$i = \iint_S N \cdot J \, dS = \text{total current passing through surface } S \quad (3)$$

These quantities are connected by four important laws discovered experimentally in the early nineteenth century. They are Faraday's law, Ampere's law, Gauss's law for electric fields, and Gauss's law for magnetic fields. We now state these four laws.

(1) Faraday's law

Faraday's law says that the integral of the tangential component of the electric intensity vector E around any closed curve C is equal to the negative of the rate of change of the magnetic flux φ passing through any surface spanning C. The equation is

$$\int_C \mathbf{E} \cdot d\mathbf{R} = -\frac{\partial \varphi}{\partial t} \tag{4}$$

(2) Ampere's law

Ampere's law says that the integral of the tangential component of the magnetic intensity vector around any closed curve C is equal to the current i flowing through any surface spanning C. The equation is

$$\int_C \mathbf{H} \cdot d\mathbf{R} = i \tag{5}$$

(3) Gauss's law for electric fields

Two quantities of charge q_1 and q_2 will attract or repel each other depending upon whether they are unlike each other or like each other. Coulomb found a remarkable law that the force of attraction or repulsion is given by the equation, known as Coulomb's law,

$$F = k \frac{q_1 q_2}{r^2}$$

where r is the distance between the two quantities q_1 and q_2 and k is a constant. The remarkable fact about Coulomb's law is that it has the same form as the law of gravitation. The charges q_1 and q_2 act like masses and the force between them varies with distance in exactly the same way as the force of gravitation between two masses. We must remember that electrical forces can be attractive or repulsive, whereas gravitational forces are always attractive.

In 1839 Karl Frederick Gauss used Coulomb's law to deduce a law for electric fields. Consider a system of particles of total charge q. Imagine that the total charge q is surrounded by a closed surface S of any shape. Such a hypothetical closed surface is called a Gaussian surface. Gauss's law is the expression

$$\iint_S \mathbf{N} \cdot \mathbf{D} \, dS = q \tag{6}$$

for the outward flux through the entire closed Gaussian surface. Equivalently we may say that Gauss's law says that the integral of the normal component of the electric flux density over any closed surface S is equal to the total electric charge q enclosed by S. Physically Gauss's law in effect says that lines of force can begin or end only on charges. Thus the conservation of lines of force is equivalent to the conservation of charge.

(4) Gauss's law for magnetic fields

The corresponding law for magnetic fields says that the total magnetic flux passing through any closed surface S is zero. The law is

$$\iint_S \mathbf{N} \cdot \mathbf{B} \, dS = 0 \qquad (7)$$

We will make use of two theorems, namely Stokes theorem and the divergence theorem. We now state these two theorems. Gauss's mathematics turned the tide in favor of Maxwell.

Stokes theorem

Stokes theorem provides an integral formula of great importance. The theorem is: If S is the portion of a regular surface bounded by the closed curve C, and if $\mathbf{F}(x, y, z)$ is a vector function possessing continuous first partial derivatives, then

$$\int_C \mathbf{F} \cdot d\mathbf{R} = \iint_S \mathbf{N} \cdot \nabla \times \mathbf{F} \, dS \qquad (8)$$

provided that the direction of integration around C is positive with respect to the side on which the unit normal N is drawn.

The divergence theorem

The divergence theorem is: If $\mathbf{F}(x, y, z)$ and $\nabla \cdot \mathbf{F}$ are continuous over the closed regular surface S and its interior V, and if $\mathbf{N}$ is the unit vector perpendicular to S at a general point and extending outward from S, then

$$\iint_S \mathbf{N} \cdot \mathbf{F} \, dS = \iiint_V \nabla \cdot \mathbf{F} \, dV \qquad (9)$$

Maxwell's equations

We will now derive the four Maxwell's equations, namely

Maxwell's first law follows from Faraday's law,

Maxwell's second law follows from Ampere's law,

Maxwell's third law follows from Gauss's law for electric fields, Maxwell's fourth law follows from Gauss's law for magnetic fields. We perform the following operations.

Maxwell's first law

We apply Stokes theorem (8) to the left hand side of Faraday law (4). We obtain

$$\int_C \mathbf{E} \cdot d\mathbf{R} = \iint_S \mathbf{N} \cdot \nabla \times \mathbf{E} \, dS \qquad (10)$$

The total magnetic flux φ passing through surface S is given by equation (1), which is

$$\varphi = \iint_S \mathbf{N} \cdot \mathbf{B} \, dS \qquad (2)$$

Thus the right hand side of Faraday law (4) becomes

$$-\frac{\partial \varphi}{\partial t} = -\frac{\partial}{\partial t}\left(\iint_S \mathbf{N} \cdot \mathbf{B} \, dS\right) = \iint_S \mathbf{N} \cdot \left(-\frac{\partial \mathbf{B}}{\partial t}\right) dS \qquad (11)$$

If we equating the right hand side of (10) to the right hand side of (11) we obtain

$$\iint_S \mathbf{N} \cdot \nabla \times \mathbf{E} \, dS = \iint_S \mathbf{N} \cdot \left(-\frac{\partial \mathbf{B}}{\partial t}\right) dS$$

Since S is an arbitrary surface spanning the arbitrary closed curve C, this equation establishes *Maxwell's first law*.

$$\nabla \times \mathbf{E} = -\frac{\partial \mathbf{B}}{\partial t} \qquad (12)$$

Maxwell's second law

Similarly, we apply Stokes theorem (8) to left hand side of Ampere's law (5). We obtain

$$\int_C \mathbf{H} \cdot d\mathbf{R} = \iint_S \mathbf{N} \cdot \nabla \times \mathbf{H} \, dS \qquad (13)$$

The total current passing through surface S is given by equation (3), which is

Chapter 10. Gauss and Maxwell's equations

$$i = \iint_S \mathbf{N} \cdot \mathbf{J}\, dS \qquad (3)$$

If we equating the right hand side of (13) to the right hand side of (3) we obtain

$$\iint_S \mathbf{N} \cdot (\nabla \times \mathbf{H})\, dS = \iint_S \mathbf{N} \cdot (\mathbf{J})\, dS$$

Since S is an arbitrary surface, we conclude that the vectors being integrated over S must be identical, that is

$$\nabla \times \mathbf{H} = \mathbf{J} \qquad (15)$$

Maxwell is immortal because he recognized that the current density $\mathbf{J}$ consists of the two parts; that is

$$\mathbf{J} = \mathbf{J}_c + \mathbf{J}_d \qquad (16)$$

One part is the conduction current density

$$\mathbf{J}_c = \sigma \mathbf{E} \qquad (17)$$

which is due to the flow of electric charge. The other part is the displacement current density

$$\mathbf{J}_d = \frac{\partial \mathbf{D}}{\partial t} = \varepsilon \frac{\partial \mathbf{E}}{\partial t} \qquad (18)$$

which is due to the time variation of the electric field. Thus

$$\mathbf{J} = \sigma \mathbf{E} + \varepsilon \frac{\partial \mathbf{E}}{\partial t} \qquad (19)$$

Equation (15) becomes *Maxwell's second law*

$$\nabla \times \mathbf{H} = \sigma \mathbf{E} + \varepsilon \frac{\partial \mathbf{E}}{\partial t} \qquad (20)$$

Maxwell's third law

We apply the divergence theorem (9) to the left hand side of Gauss's law for (6) for electric fields. We get

$$\iint_S \mathbf{N} \cdot \mathbf{D}\, dS = \iiint_V \nabla \cdot \mathbf{D}\, dV \qquad (21)$$

Thus Gauss's law for (6) becomes

$$\iiint_V \nabla \cdot \mathbf{D} \, dV = q \qquad (22)$$

The total charge within volume V is given by (1), which is

$$q = \iiint_V \rho \, dV \qquad (1)$$

Thus equation (22) becomes

$$\iiint_V \nabla \cdot \mathbf{D} \, dV = \iiint_V \rho \, dV \qquad (23)$$

Since V is arbitrary, this equation establishes *Maxwell's third law*

$$\nabla \cdot \mathbf{D} = \rho \qquad (24)$$

Maxwell's fourth law.

We apply the divergence theorem (9) to the left hand side of Gauss's law (7) for magnetic fields, which is

$$\iint_S \mathbf{N} \cdot \mathbf{B} \, dS = 0 \qquad (7)$$

We obtain

$$\iint_S \mathbf{N} \cdot \mathbf{B} \, dS = \iiint_V \nabla \cdot \mathbf{B} \, dV \qquad (25)$$

From (7) and (25) we have

$$\iiint_V \nabla \cdot \mathbf{B} \, dV = 0$$

Since V is arbitrary, this equation establishes *Maxwell's fourth law*.

$$\nabla \cdot \mathbf{B} = 0 \qquad (26)$$

In summary, Maxwell's Equations are

$$\nabla \times \mathbf{E} = -\frac{\partial \mathbf{B}}{\partial t} \qquad \nabla \cdot \mathbf{D} = \rho$$

$$\nabla \times \mathbf{H} = \sigma \mathbf{E} + \varepsilon \frac{\partial \mathbf{E}}{\partial t} \qquad \nabla \cdot \mathbf{B} = 0$$

www.ingramcontent.com/pod-product-compliance
Lightning Source LLC
Chambersburg PA
CBHW071420170526
45165CB00001B/345